GRADE 1

PROBLEM SOLVED: BAR MODEL MATH

Bob Krech

New York • Toronto • London • Auckland • Sydney
Mexico City • New Delhi • Hong Kong • Buenos Aires

Editor: Maria L. Chang
Cover design by Tannaz Fassihi
Cover art by Matt Rousell
Interior design by Grafica Inc.
Interior illustrations by Mike Moran

ISBN: 978-0-545-84009-5

Printed in the U.S.A.
First printing, June 2016.

1 2 3 4 5 6 7 8 9 10 40 22 21 20 19 18 17 16

Table of Contents

Introduction 4

Chapter 1: Using Pictures and Unit Squares With Addition Problems
- Lesson 1: Addition Using Pictures 8
- Lesson 2: Addition Using Pictures and Unit Squares 11
- Lesson 3: Addition Using Unit Squares 13
- Lesson 4: Drawing Unit Squares to Add 15

Chapter 2: Using Pictures and Unit Squares With Subtraction Problems
- Lesson 5: "Take Away" Problems Using Pictures and Unit Squares 18
- Lesson 6: "Comparison" Problems Using Pictures and Unit Squares 20
- Lesson 7: Subtraction Using Unit Squares 22
- Lesson 8: Drawing Unit Squares to Subtract 24

Chapter 3: Using Connected Squares With Addition and Subtraction (Within 20)
- Lesson 9: Addition With Larger Numbers 27
- Lesson 10: Addition With Three Addends 30
- Lesson 11: Subtraction With Larger Numbers 32
- Lesson 12: More Subtraction With Larger Numbers 34

Chapter 4: Using Connected Squares With Measurement Problems
- Lesson 13: Measurement Addition Problems. 37
- Lesson 14: Measurement Subtraction Problems 39

Chapter 5: Using Connected Squares With Money Problems
- Lesson 15: Money Addition Problems 42
- Lesson 16: Money Subtraction Problems. 44

Chapter 6: Bar Modeling With Multistep Problems
- Lesson 17: Mixed Multistep Problems 46
- Lesson 18: More Mixed Multistep Problems 49
- Lesson 19: Mixed Multistep Problems With Larger Numbers 52
- Lesson 20: Challenging Multistep Problems 54

Reproducible Word Problems 56

Answer Key. 96

Introduction

The very first process standard outlined by both the National Council of Teachers of Mathematics (NCTM) and the more recent Common Core State Standards (CCSS) focuses on understanding problems and "persevering in solving them." They purposefully listed this standard first because problem solving is what teaching math is all about. All of the skills, concepts, knowledge, and strategies children learn in math are basically tools. The eventual goal is that students will apply these math tools to solve problems they encounter in life.

Problem Solved: Bar Model Math is designed to help you and your students learn about a new problem-solving tool—a versatile and effective strategy commonly known as Bar Modeling. A major component of Singapore Math, Bar Modeling has been proven effective in helping students achieve high levels of mathematical competency.

For more than two decades, Singapore's students have consistently ranked among the highest in the world in international math assessments, such as TIMSS (Trends in International Mathematics and Science Study) and PISA (Programme for International Student Assessment). Looking at this success, many schools and districts in the United States and around the world have begun to examine and use ideas found in the Singapore Math curriculum. In my opinion, Bar Modeling is one of the most powerful—if not *the* most powerful—component of Singapore Math.

What Is Bar Modeling?

Bar Modeling is a unique and incredibly versatile strategy that can be used effectively by the youngest elementary school children all the way through to college math majors. This strategy can be applied to a wide range of problem types and contexts. We typically teach children many different strategies (such as Look for a Pattern, Draw a Picture, or Guess and Check) for tackling word problems. With the Bar Modeling method, we can attack any of the various problem types with one singular, powerful approach.

In Bar Modeling, we break out the information in a problem and represent it in a simplified pictorial form using unit squares or bars to represent quantities. The pictorial representation helps children better see and understand the quantities in—and thus the possible solutions to—the problem. The great advantage of the method is that children can visually represent both the given facts in the problem as well as the "unknown" (what they are trying to find out) in a way that allows them to *see* relationships between the quantities, thus promoting flexible thinking about numbers and general number sense. Bar Modeling empowers children to become thinkers, not memorizers.

Where Does Bar Modeling Fit Into the Problem-Solving Process?

When solving word problems, it is always helpful to guide children to using a consistent process. A typical five-step process includes the following:

1. Identify and underline the facts.
2. Identify and circle the question.
3. Eliminate any unnecessary information.
4. Choose a strategy and solve.
5. Go back and check if the answer makes sense.

Bar Modeling falls under the "Choose a strategy and solve" step.

How Does Bar Modeling Work?

When first learning how to solve word problems, children find it very helpful to have problems represented by physical objects. For example, an appropriate problem for primary grades might be: "*Sara has 2 apples. Dana has 4 apples. How many apples do the girls have altogether?*" Instead of using actual apples, children might use physical manipulatives, such as cubes or blocks, to represent the apples. Later, children might draw apples or even squares to make the representation a little more abstract. Bar Modeling takes the pictorial representation a step further toward the abstract. Instead of drawing apples or squares, children represent the quantities by drawing simple bars or rectangles labeled with words, arrows, and numbers. To solve a problem about ladybugs in a garden, for example, children's diagrams would progress as below:

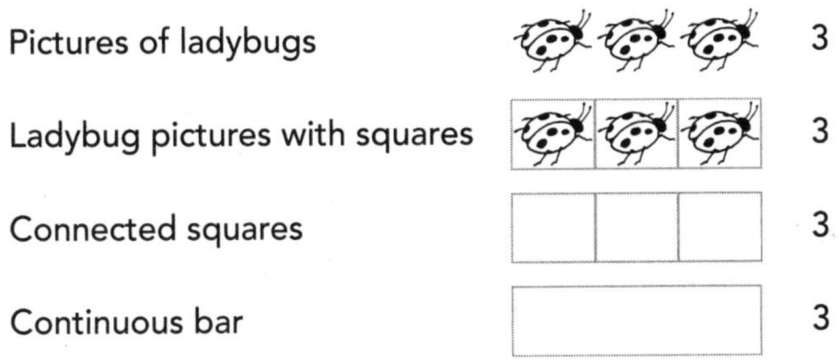

Pictures of ladybugs 3

Ladybug pictures with squares 3

Connected squares 3

Continuous bar 3

This book lays the foundation for Bar Modeling even though it doesn't teach 1st graders how to draw continuous bar models, which may be too abstract for them. Instead, the book shows children step-by-step how to draw pictures or connected squares in such a way that helps them understand what a problem is asking and see the relationships between the amounts given. They learn how to label their diagrams properly and how to use an arrow and question mark to indicate the unknown quantity for which they are solving. Children also discover strategies for finding the unknown, using their diagrams effectively. Eventually when children have grown comfortable using connected squares to represent the problems, they can erase the lines between the squares to reveal a continuous bar. The transition to actual Bar Modeling will then be easy, since they already know the basics of it.

Bar Modeling Variations

The idea of modeling a problem using connected squares or continuous bars can be applied to a wide range of math problems. Within Bar Modeling, there are variations that suit different types of problems. All variations can represent quantities of anything (e.g., apples, miles, dollars, minutes) using labeled bars. (NOTE: From here on, the term *bar* could include connected squares.)

Part-Whole Model

In the Part-Whole Model, two or more parts make up the whole. This model works well for subtraction involving "take away" problems. We can represent the operation of subtraction by crossing out or shading in the quantity being "taken away." The remaining section, or "unknown," is indicated by an arrow and a question mark.

Example: *Priya has 8 baseball cards. She gave 5 cards to Stan. How many baseball cards does Priya have left?*

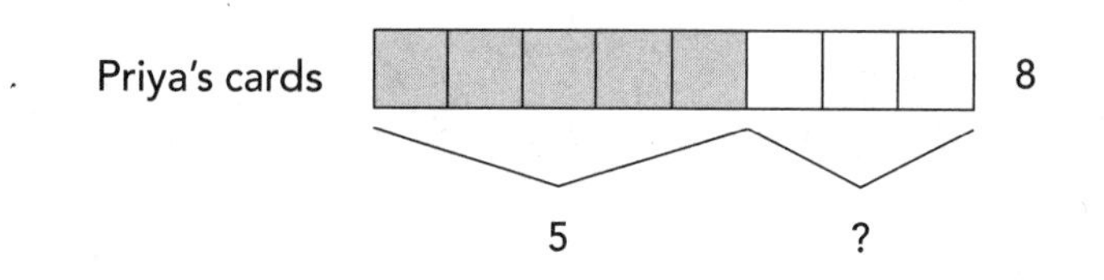

Comparison Model

In the Comparison Model, each quantity in a problem is represented by its own bar. This model has certain advantages when looking closely at relationships between numbers and when breaking up numbers (by place values, for example), so it works particularly well for addition and "comparison" problems.

Addition Example: *Stan has 5 baseball cards. Priya has 8 baseball cards. How many baseball cards do they have altogether?*

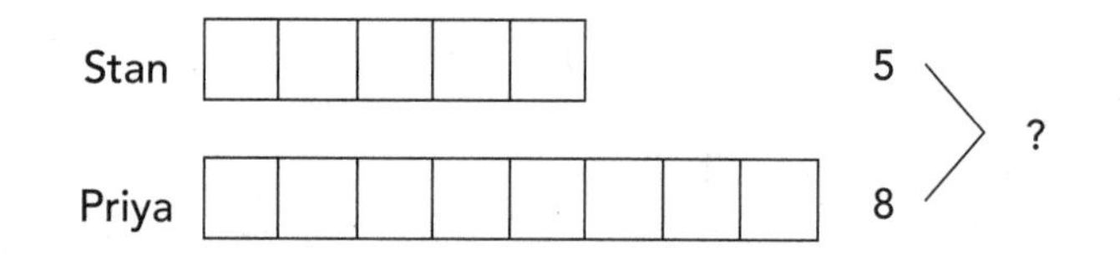

The arrow connecting the two numbers shows that the two quantities should be joined or added together to find the unknown, which is indicated by the question mark.

Graphing or Lined Paper

When children work on problems independently (other than the ones included in this book), it might be helpful to provide them with graph paper or lined paper. Turned horizontally, these types of paper provide lines that children can use as guides for drawing squares and bars.

Comparison Example: *Priya has 8 baseball cards. Stan has 5 baseball cards. How many more baseball cards does Priya have than Stan?*

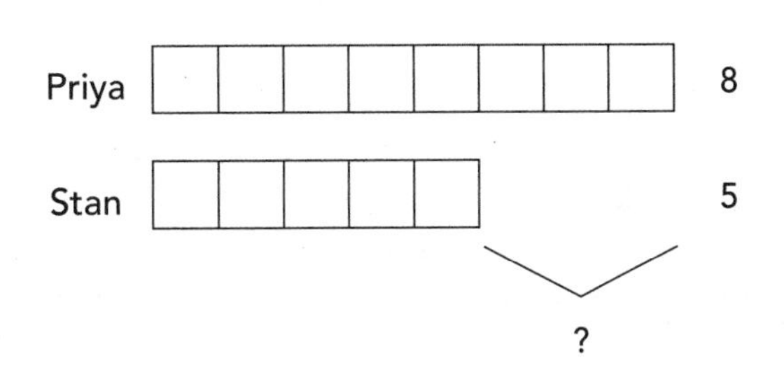

Here, the two bars representing the quantities can be compared easily. The difference between the bars, which is the unknown in this problem, is indicated by the arrow and question mark.

The names of the various models are not so important. Different books and teachers use different names. As you work through the problems in this book, you will find there is much overlap among the methods and there's room for you and your students to create your own variations. Each lesson will detail the actual step-by-step procedures for drawing bar models for these basic variations, particularly in the first two chapters, which are on addition and subtraction.

What Is in This Book?

This book offers a step-by-step approach to helping children learn the Bar Modeling strategy in the context of solving word problems. There are 20 lessons, each with four carefully chosen word problems that gradually increase in difficulty. Use the first two problems in each lesson to introduce the strategy, then have children practice it on their own with the other two problems. All of the word problems are presented in two ways:

- A digital version can be accessed through **www.scholastic.com/problemsolvedgr1** (registration is required). Display the problems on the interactive whiteboard during your lesson to support your teaching.
- A reproducible version for children can be found in the second part of this book, starting on page 56. These pages feature a lightly printed graph-paper background to help children keep their diagrams neat and organized.

As you work with children on the various word problems, follow the problem-solving process outlined at the top of page 5. Repeating this routine regularly will help children confidently tackle any problem they encounter, especially in testing situations. By the time you have completed the lessons in this book, you will have equipped your students with another powerful tool they can count on as they continue to develop their abilities as excellent, world-class problem solvers.

Silly Problems to Engage Children

The story problems in this book include characters and situations guaranteed to amuse elementary grade children. Silly names (like Willy Worm) and ridiculous situations (like a teacher who finds raccoons in her students' desks) tend to capture children's interest. Studies have shown that when humor is engaged, attention is engaged. Children will enjoy reading these problems and look forward to solving them.

Using Pictures and Unit Squares With Addition Problems

As an introduction to the concept of Bar Modeling, children will start by representing the quantities in a word problem with rows of pictures then transition to drawing unit squares. They will label the diagrams with a number at the end of each row to indicate the quantity, and a word or initial in front of each row to identify what the quantity represents. A simple arrow with a question mark will indicate the unknown quantity in the problem.

LESSON 1

Addition Using Pictures

Materials: student pages 56–57, pencils, projector, interactive whiteboard, markers

Preparation: Distribute copies of pages 56–57 and pencils to children. Go to www.scholastic.com/problemsolvedgr1 and click on Lesson 1. Set up your computer and projector to display the problems on the interactive whiteboard.

Display Problem #1 on the interactive whiteboard.

Flip found 3 stinkbugs. Flam found 4 stinkbugs.
How many stinkbugs did they find in all?

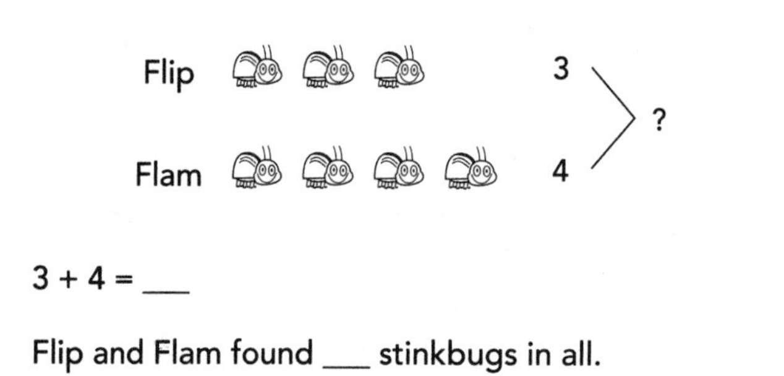

3 + 4 = ___

Flip and Flam found ___ stinkbugs in all.

Say: *Today we will begin learning about a way to solve word problems. It is called Bar Modeling. It will use some ideas we already know about math and solving problems. But we will also learn some new ideas. Let's take a look at a problem.* Read aloud the problem on the board.

Ask: *What is the problem telling us?* (Flip found 3 stinkbugs. Flam found 4 stinkbugs.) Underline these facts on the board and have children do the same on their papers. Point to the drawings on the board.

Say: *Here we have drawings of Flip's 3 stinkbugs and Flam's 4 stinkbugs. Flip's is in one row, and Flam's is on another row right below it.*

Ask: *How do we know which row is Flip's and which is Flam's?* (Their names are at the front of the rows.) As children respond to the question, highlight the names on the board.

Ask: *How do we know how many stinkbugs are in each row?* (We can count them or look at the number at the end of each row.) Highlight the numbers on the board.

Ask: *What else do you notice at the end of the rows?* (An arrow and a question mark) Highlight the arrow and question mark on the board.

Explain: *That arrow joins the two numbers together. It tells us to find out what those two numbers add up to. We call the question mark the* unknown, *because we don't know what it is yet. We might also call it the* answer, *or since we're adding, we can also call it the* sum.

Ask: *What are we trying to find out?* (How many stinkbugs did they find in all?) Circle the question on the board and have children do the same on their papers.

Say: *So the question is, how many stinkbugs did they find in all. That is our unknown.*

Point to the answer sentence at the bottom of the board and read it aloud: "Flip and Flam found *blank* stinkbugs in all."

Explain: *This sentence tells us what we are trying to find out, using words. We will write our answer in the blank in this sentence.* Highlight the answer sentence on the board.

Ask: *What else do you see on the board?* (A number sentence: 3 + 4 = __)

Say: *This number sentence is another way to write what we are doing and trying to find out. But instead of words, we are using numbers. We are adding Flip's 3 stinkbugs and Flam's 4 stinkbugs to find out how many there are in all. This addition sentence, 3 + 4, shows that we are adding them together.* Highlight the number sentence on the board.

Have children work in pairs to solve the problem and fill in the blanks on the number sentence and word sentence on their papers.

Invite children to share their strategies and answers with the class. There are many ways to solve this problem. Some children may just count each of the 7 pictures one-by-one. Some may see that they could count by 2s and add on 1 more. As children share their strategies, diagram their thinking on the whiteboard. For example, if a child says, "I counted by 2s and added 1 more," you could illustrate that on the board (see below).

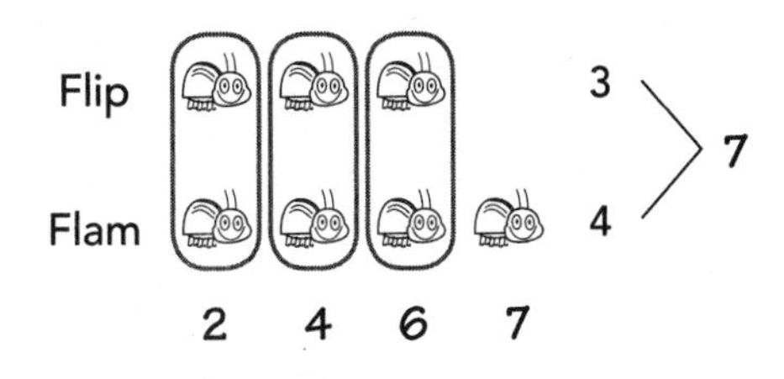

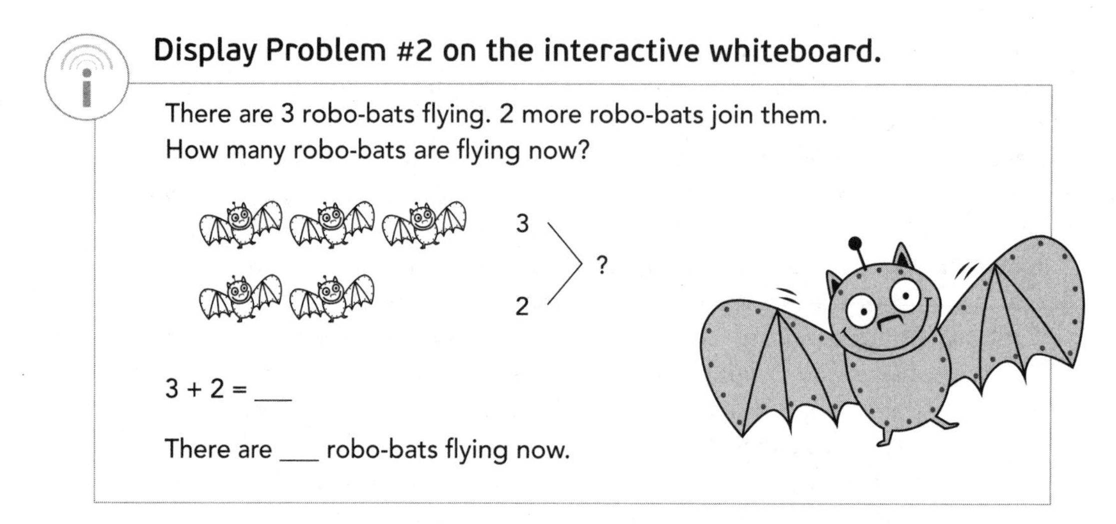

Say: *Let's try another problem.* Read aloud the problem. Have children identify the facts (3 robo-bats are flying and 2 more join them) and underline them on their papers. Then have them identify the question and circle it on their papers. Read aloud the answer sentence: "There are *blank* robo-bats flying now."

Say: *Again we have drawings of the robo-bats. They are drawn in two rows. This is a good way to organize the two amounts. It is easier to count and add them than if they were all in one long row. The top row shows the first group of robo-bats flying. The second row shows the robo-bats that joined them.*

Point to the arrow and question mark at the end of the rows.

Say: *This arrow tells us the two numbers should be joined together. The question mark tells us that is what we're trying to find out—our unknown.*

Ask: *How can we find out how many robo-bats are flying?*

Invite children to suggest some strategies, such as counting up 2 from 3 or finding 2 + 2 and adding 1 more. Discuss these ideas and diagram them on the board by circling or outlining quantities.

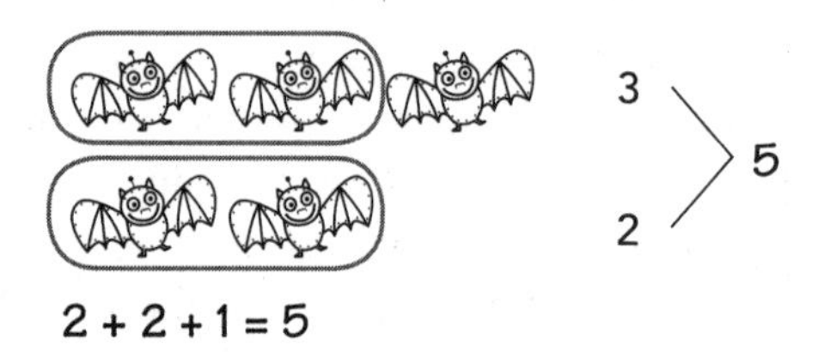

Conclude by having children fill in the answer blank and completing the number sentence.

Have children work on Problems #3 and 4 (page 57) in pairs. Give them a few minutes to work. Then display each problem on the whiteboard. Invite children to share their strategies for solving the problems, interacting on the whiteboard whenever possible.

LESSON 2

Addition Using Pictures and Unit Squares

Materials: student pages 58–59, pencils, projector, interactive whiteboard, markers

Preparation: Distribute copies of pages 58–59 and pencils to children. Go to www.scholastic.com/problemsolvedgr1 and click on Lesson 2. Set up your computer and projector to display the problems on the interactive whiteboard.

This lesson transitions from pictures to unit squares as a way to represent quantities. In the whiteboard problems, the pictures will first appear as they did in Lesson 1 and then they will appear with a square drawn around each picture. Each of these squares is a discrete unit called a *unit square*. Student worksheets will display this same discrete unit format. The diagrams will be labeled and an arrow and question mark will indicate the unknown quantity.

Display Problem #5 on the interactive whiteboard.

Maxie ate 2 super cookies. His sister Flaxie ate 5 super cookies. How many super cookies did they eat altogether?

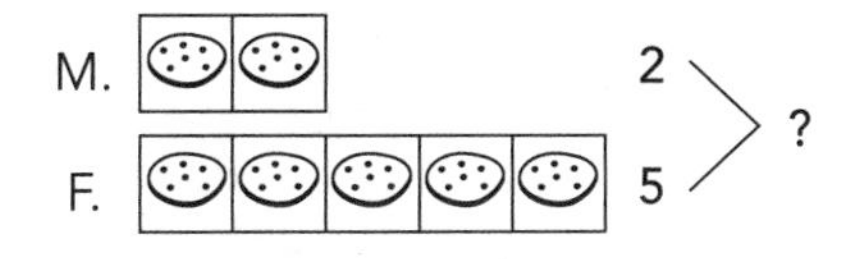

2 + 5 = ___

Maxie and Flaxie ate ___ super cookies altogether.

Read aloud the problem. Start the problem-solving process by having children identify the facts (Maxie ate 2 super cookies and Flaxie ate 5 super cookies) and underline them on their papers. Then have them circle the question as well.

Say: *Just like in our last lesson, we have drawings of the cookies in this problem.*

Click on the next page to display another version of the picture, with a square drawn around each cookie.

Say: *Now each cookie has a square drawn around it. See how the squares are connected in a row? This makes it easier to count and add the cookies.*

Point to the labels *M* and *F* in front of the rows.

Ask: *What do you think* M *and* F *stand for?* (Maxie and Flaxie)

Explain: *We can use an abbreviation or letter to say what each row is for. We don't always have to write out the whole word.*

Invite children to suggest strategies for solving the problem. Common strategies include counting up from the larger number, in this case, 5. You might want to tell children that the pictures showing the quantities don't have to be drawn in the same order they appear in the problem. In this problem, for example, it might be easier to see the row of 5 above the row of 2 (see the diagram on page 12). This would make for a more natural counting-up situation.

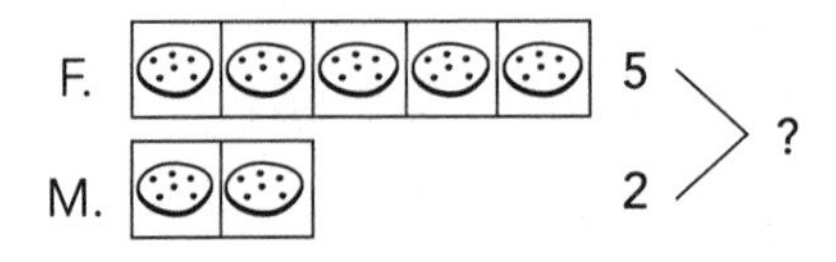

In addition to counting up, children might see a 3 + 3 double that makes 6 and then adding 1 more provides the answer of 7. Some children might also see an opportunity to count by 2s. Share the diagram below on the board to show how outlining or circling these groupings might look.

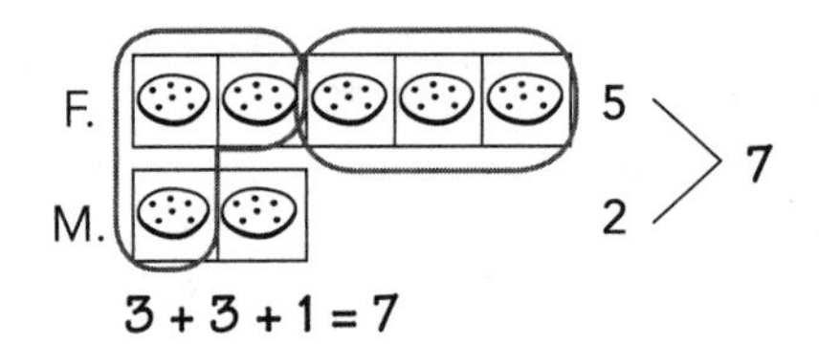

Display Problem #6 on the interactive whiteboard.

There were 6 birds sitting in a nest. Then 1 more bird joined them. How many birds are in the nest now?

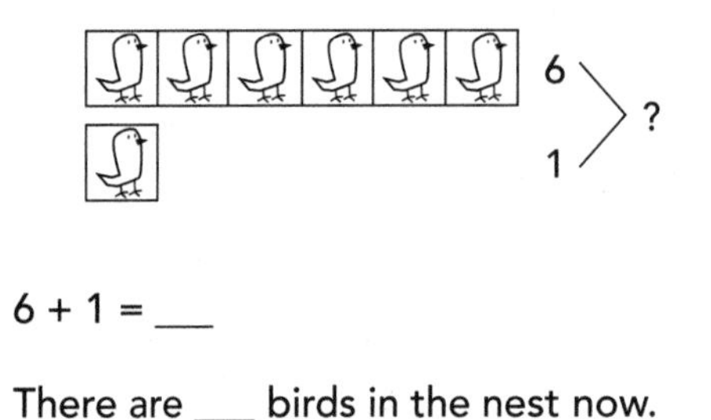

6 + 1 = ___

There are ___ birds in the nest now.

Labeling

Point out to children that we should label quantities in a diagram only when it is helpful. Don't make labeling a requirement for every problem. For example, if a problem compares quantities of carrots and peas, it makes sense to label each quantity to keep them straight. However, if a problem talks about a quantity of pencils, in which some were added or taken away, there's no need to label that quantity since there is nothing to differentiate.

Read aloud the problem. Have children identify the facts and underline them on their papers. Have them circle the question as well.

Say: *Here again we have drawings of birds to help us see the problem.* Click on the next page to display another version of the picture on the board.

Say: *Now each bird has a square around it. This makes it easier for us to count and add the birds.*

Invite children to share their strategies for solving the problem. The most common strategy here will be to count up 1 from 6. The diagram lends itself to this solution. It is also possible to count by 2s or even by 3s, then adding on 1 more to get to 7.

Point out to children that this sum (7) is the same as the sum in Problem #5. Engage children in a discussion about how this is possible given that the addends are different. Challenge them to think of other ways to make 7.

Have children work on Problems #7 and 8 (page 59) in pairs. Give them a few minutes to work. Then display each problem on the whiteboard. Invite children to share their strategies for solving the problems, interacting on the whiteboard whenever possible.

LESSON 3

Addition Using Unit Squares

Materials: student pages 60–61, pencils, projector, interactive whiteboard, markers
Preparation: Distribute copies of pages 60–61 and pencils to children. Go to www.scholastic.com/problemsolvedgr1 and click on Lesson 3. Set up your computer and projector to display the problems on the interactive whiteboard.

Children will start drawing unit squares around pictures themselves to create discrete units that are easier to count and add. They will also begin to label the diagrams with numbers and names (or abbreviations) where helpful, and an arrow and question mark to indicate the unknown quantity.

Display Problem #9 on the interactive whiteboard.

Stan caught 5 Giganto-Fish. Fran also caught 5 Giganto-Fish.
How many Giganto-Fish did they catch in all?

S. [5 fish in squares] 5
F. [5 fish in squares] 5
?

5 + 5 = ___

Stan and Fran caught ___ Giganto-Fish in all.

Begin as always by reading aloud the problem. Help children get in the habit of using the problem-solving process. Have them identify the facts (Stan caught 5 Giganto-Fish; Fran also caught 5 Giganto-Fish) and underline them on their papers. Then have them circle the question as well.
Say: *In our last lesson, we had pictures of the things in each word problem. Then we saw a square drawn around each picture. We saw how the squares helped keep the things organized and easy to count. For example, when we had a story problem about cookies, there was a square around each cookie. Today, we will start again with pictures. But this time you will draw the squares around them.*

Model how to draw a square around each fish on the board, and have children do the same on their papers.
Say: *Notice the gray lines on your paper. They are there to help you draw squares neatly. The squares will help us count and keep track of each fish in a careful way so we don't miss any.*

With the class, count the squares with the Giganto-Fish..Point out that the number at the end of each row matches the number of squares. Then bring children's attention to the letters S and *F* at the beginning of the rows.
Ask: *Why do you think these rows are labeled* S *and* F? (To show that one row is for Stan and one row is for Fran)

Next, point to the arrow and question mark.

Ask: *What do the arrow and question mark tell us?* (They tell us that we need to join the two numbers to find the unknown or answer.)

Solve the problem together, encouraging children to share their strategies. Some children may know the 5 + 5 doubles fact, while others may count by 2s in the vertical columns to get to 10.

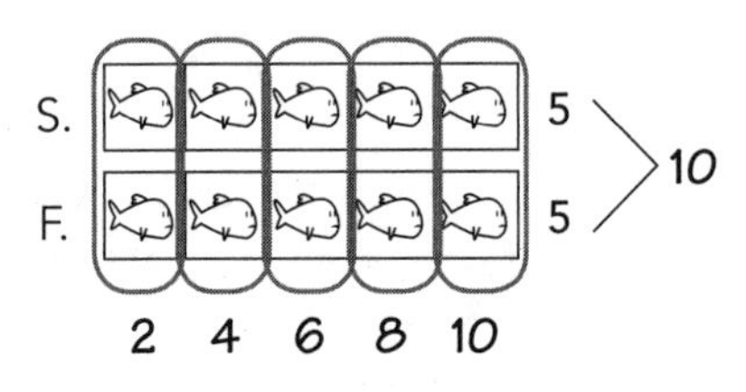

Display Problem #10 on the interactive whiteboard.

Billy Worm counted 4 birds. Willy Worm counted 5 birds.
How many birds did they count altogether?

4 + 5 = ___

They counted ___ birds altogether.

Read aloud the problem. Have children identify and underline the facts and circle the question on their papers.

Say: *Again we have pictures that show how many birds there are. But this time there are no numbers, labels, or unknowns. We will add these labels to our diagrams. Let's start by drawing a square around each bird to make the birds easier to count. Remember, you can use the gray lines on your paper to help you draw your squares neatly.* Model this on the board.

Say: *Let's count how many squares are in each row. Then we'll write the number at the end of each row. Make sure to line up those numbers, one above the other, even though the rows are not the same length.*

Continue: *Now let's label the rows. How should we do that?* (Write *B* for Billy and *W* for Willy.) Model this on the board and have children do the same on their papers.

Say: *Now we have to show where our unknown is. How do we do that?* (Draw an arrow joining the two numbers and write a question mark.) Do this on the board and check that children have done this on their papers as well.

Solve the problem together, encouraging children to share their strategies. Helpful strategies include finding and outlining the 4 + 4 double, as well as counting by 2s and adding 1 more.

Have children work on Problems #11 and 12 (page 61) in pairs. Remind them to draw unit squares around the pictures and to label the diagrams. Give children a few minutes to work. Then display each problem on the whiteboard. Call on volunteers to label the diagrams on the board. Then invite children to share their strategies for solving the problems, interacting on the whiteboard whenever possible.

LESSON 4

Drawing Unit Squares to Add

Materials: student pages 62–63, pencils, projector, interactive whiteboard, markers
Preparation: Distribute copies of pages 62–63 and pencils to children. Go to www.scholastic.com/problemsolvedgr1 and click on Lesson 4. Set up your computer and projector to display the problems on the interactive whiteboard.

At the beginning of this lesson, children will review how pictures can be organized and counted more easily when each one is surrounded by a square. They will then see that squares without pictures inside can still represent the quantities in a problem effectively. Children will read the problem and draw their own squares to represent the quantities. They will also label their diagrams with numbers, words when helpful, and an arrow and question mark to indicate the unknown quantity.

Display Problem #13 on the interactive whiteboard.

Begin as always by reading aloud the problem. Have children start the problem-solving process by underlining the facts and circling the question on their papers. Point out how each sock on the board has a square around it. Invite children to count the socks with you. Then click on the next page to reveal another version of the diagram that shows just the squares without the socks. Count the squares again.

"Represent"— A Key Word

Throughout this book we tell children to "represent" quantities with drawings and symbols. This is a major idea in math—we don't always need the physical objects to count, add, subtract, or do any other operation. Instead we can use drawings or symbols to *represent*, or stand for, amounts. Tell children: *In math we often use a simple drawing or symbol to represent, or stand for, something that might be too hard or take too long to draw. A square can represent a cat, a dog, a cookie, a rocket—anything you want.*

Say: *Drawing a square is much easier and quicker than drawing a sock or a fish or a person or even a raccoon. On your paper, draw a square for each sock, just like you see on the board. The squares do not have to be perfect. You do not need a ruler. Just use the gray lines on your paper to help you draw a square neatly. Also, remember to make a row for each amount in the problem. Make sure the squares are lined up one above the other.* Give children a few minutes to complete their drawings.
Ask: *How many rows of squares did you draw?* (2) *How many squares did you draw for Pete?* (5) *How many for Paul?* (4)
Say: *Now let's label the diagram with names, numbers, and an arrow and question mark. How should you label the rows?* (One row for Pete and the other row for Paul. Pete's row should be labeled 5 and Paul's should be labeled 4.) Call on volunteers to label the diagram on the board.

Have children work in pairs to share strategies and solve the problem. Strategies might include looking for doubles, such as 3 + 3 or 4 + 4. Counting by 2s then adding 1 more also makes sense. Encourage children to circle the doubles or the 2s in their diagram to help them see the strategies at work.

Display Problem #14 on the interactive whiteboard.

Pinky the poodle has 7 blue dog collars and 3 green dog collars. How many dog collars does Pinky have in all?

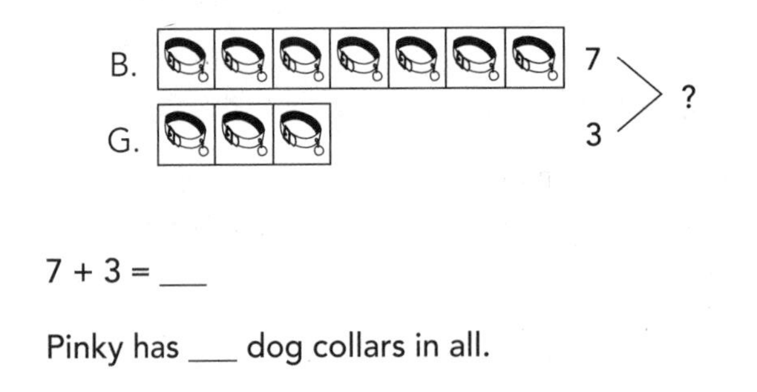

7 + 3 = ___

Pinky has ___ dog collars in all.

Read aloud the problem. Then have children underline the facts and circle the question on their papers.
Say: *Here is the problem with the dog collars drawn with squares around them.* Click on the next page to display another version of the diagram that shows just the squares without the collars.
Say: *Now we see only the squares. On your paper, draw a square for each dog collar, just like you see on the board. Remember to make a row for each amount in the problem. Make sure the squares are lined up one above the other.* Give children a few minutes to complete their drawings.
Say: *Now let's label the diagram with names, numbers, and an arrow and question mark. Remember, when labeling the rows we don't have to use the full word, like* blue *or* green. *We can just use a letter like* B *or* G.

Check to make sure children drew and labeled their diagrams correctly, then call on volunteers to draw and label the diagram on the board.

Engage children in a discussion about various strategies for solving the problem. Some children may start at 7 and count up 3 to 10. Others may look for the 3 + 3 double and count up 4 more or see three 3s and add 1 more.

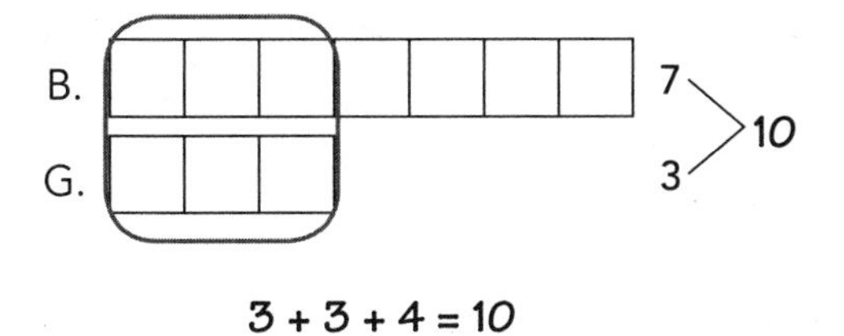

Have children work on Problems #15 and 16 (page 63) in pairs. Remind them to draw unit squares to represent the amounts in each problem and to label them. Give children a few minutes to work. Then display each problem on the whiteboard. Call on volunteers to draw and label the diagrams on the board. Then invite children to share their strategies for solving the problems, interacting on the whiteboard whenever possible.

Homework

If you want children to practice these strategies at home, assign developmentally appropriate word problems from any source. Ask children to solve the problem using the Bar Modeling method that you have been practicing in class. Provide graph paper or horizontally lined paper to help children with their drawings.

CHAPTER 2

Using Pictures and Unit Squares With Subtraction Problems

Subtraction is about finding the difference between two quantities. Sometimes a problem begins with a quantity and some are taken away from that. For example: *Bob had 4 cookies. He ate 2 cookies. How many cookies are left?* Other times a problem compares two quantities: *Karen had 6 cookies. Bob had 4 cookies. How many more cookies did Karen have than Bob?*

Subtracting and adding up are both reasonable strategies for finding the answer, but either way children are finding the difference between the two quantities. Bar Modeling is a great way to help children visualize, represent, and understand subtraction—in both "take away" and "comparison" situations.

Get children in the habit of going through the problem-solving process with each problem they encounter: read aloud the problem, identify and underline the facts (what we know), and circle the question. As much as possible, have children work in pairs.

LESSON 5

"Take Away" Problems Using Pictures and Unit Squares

Materials: student pages 64–65, pencils, projector, interactive whiteboard, markers

Preparation: Distribute copies of pages 64–65 and pencils to children. Go to www.scholastic.com/problemsolvedgr1 and click on Lesson 5. Set up your computer and projector to display the problems on the interactive whiteboard.

This lesson focuses on basic subtraction problems. As with addition, the process begins with simple drawings to represent the quantities in the problem. Then the drawings will appear with a square placed around each picture, making the unit more discrete.

Display Problem #17 on the interactive whiteboard.

Zip had 4 moon rocks. He threw 2 moon rocks at the Goo Goo Monster. How many moon rocks did Zip have left?

4

?

4 – 2 = ___

Zip had ___ moon rocks left.

Ask: *What facts do we know in this problem?* (Zip had 4 moon rocks. He threw 2 moon rocks at the Goo Goo Monster.) Underline the facts on the board and have children do the same on their papers.

Say: *On your page, you will notice all of the moon rocks are drawn in one row. There is a number at the end of the row that says how many rocks are there. There are also some moon rocks crossed out.*

Ask: *Where are the arrow and question mark pointing to in this problem?* (To the moon rocks that have not been thrown) *How is this different from the addition problems?* (We are not joining things together. We are trying to find out what is left when some things are taken away.)

Next, ask children to identify the question and circle it on their papers. Then have them find the answer sentence and read it together: "Zip had *blank* moon rocks left."

Ask: *What are we trying to find out?* (How many moon rocks are left?) *How can we show this on our diagram?* (Cross out the moon rocks that Zip threw.)

On the interactive whiteboard, model crossing out the 2 moon rocks that were thrown. Simply draw over the work already done.

Say: *We can see there are two rocks that are not crossed out. These are the rocks that are left. We see the arrow with the question mark showing that this is the unknown quantity.*

Encourage children to share strategies and check their answers. Have them fill in the blank in the answer sentence and in the number sentence, or equation.

Crossing Out—Does Direction Matter?

When subtracting, explain to children that it doesn't matter where they begin crossing out or shading in the row of pictures or squares. Starting at the front or back of the row will yield the same results. You may want to demonstrate this a couple of times to show that the results are the same. Then let children use their personal preference.

Display Problem #18 on the interactive whiteboard.

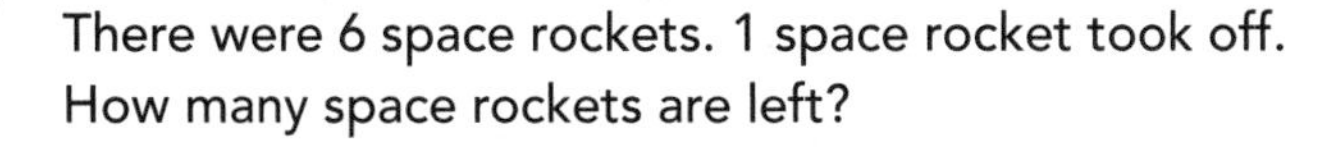

There were 6 space rockets. 1 space rocket took off.
How many space rockets are left?

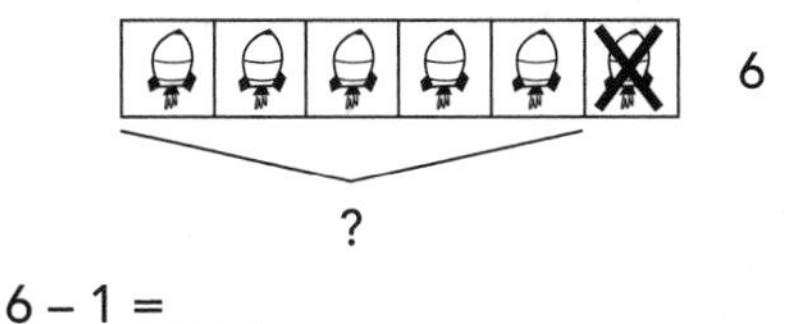

$6 - 1 =$ ___

There are ___ space rockets left.

Read aloud the problem. Have children underline the facts and circle the question on their papers. Read the answer sentence together aloud: "There are *blank* space rockets left."

Say: *Just like in our last problem, we have drawings of the rockets in this problem.* Click on the next page to display another version of the picture that shows a square around each rocket.

Say: *Now there's a square around each rocket. That makes them easier to count.*

Ask: *How many space rockets are there at the start of the problem?* (6) *How many rockets took off?* (1) *How can we show that one is gone?* (Cross out or shade in the square.)

Model crossing out one space rocket on the board and have children do the same on their papers.

Ask: *How do we know what is left?* (We can count the rockets that are not crossed out.)

Guide children to notice the unknown, as shown by the arrow and question mark pointed at the remaining rockets.

Say: *Here is our unknown—the number of rockets left.*

There are various strategies for efficiently counting the remaining rockets, a favorite being counting by 2s. Have children share their strategies. Conclude by having them fill in the blanks in the answer sentence and in the equation.

Have children work on Problems #19 and 20 (page 65) in pairs. Remind them to draw unit squares around the pictures and to cross out items as they subtract. Give children a few minutes to work. Then display each problem on the whiteboard. Invite children to share their strategies for solving the problems, interacting on the whiteboard whenever possible.

LESSON 6

"Comparison" Problems Using Pictures and Unit Squares

Materials: student pages 66–67, pencils, projector, interactive whiteboard, markers

Preparation: Distribute copies of pages 66–67 and pencils to children. Go to www.scholastic.com/problemsolvedgr1 and click on Lesson 6. Set up your computer and projector to display the problems on the interactive whiteboard.

This lesson focuses on "comparison" subtraction problems. Children will continue to work with drawings that have unit squares around them.

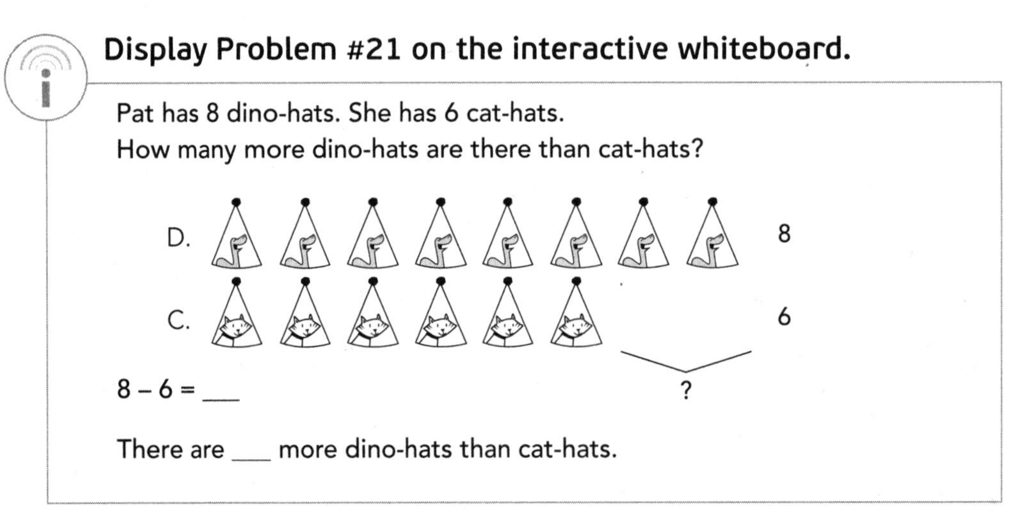

Read aloud the problem. Have children underline the facts and circle the question on their papers.

Explain: *Here is our first comparison problem. Nothing is being taken away in*

this problem. Instead we have two different things: cat-hats and dino-hats. The picture shows two rows of hats, one above the other, just like when we were adding. But this time, we're not trying to find out how many hats there are in all.
Ask: *What is the question asking?* (How many more dino-hats are there than cat-hats?) *So which are there more of—dino-hats or cat-hats?* (Dino-hats)
Say: *We can see in the picture that there are more dino-hats. What we need to find out is, how many more? So we need to compare the two amounts. To do that, we use subtraction.*

Guide children to notice the arrow pointing to the area at the end of the second row. The question mark indicates two ideas simultaneously: It shows there are 2 less cat-hats, and there are 2 more dino-hats. Draw a line at the end of the 6 cat-hats through the row of dino-hats, as shown below, to emphasize this point. As you demonstrate this, tell children how drawing the line helps us see the difference between the two quantities.

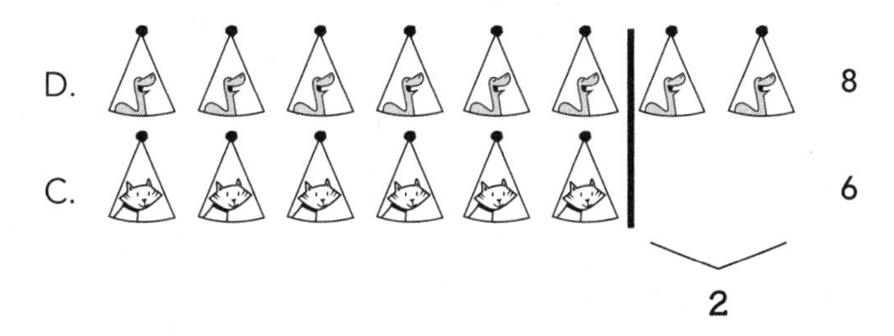

Ask children to find an answer to the problem, share their strategies with partners, and then share back solutions with the class.

Some solutions might include connecting dino-hats and cat-hats in one-to-one correspondence with lines or just counting the leftover dino-hats that don't match up with cat-hats. In either case, help children understand that this is finding the difference and we can express it with a subtraction sentence.

Display Problem #22 on the interactive whiteboard.

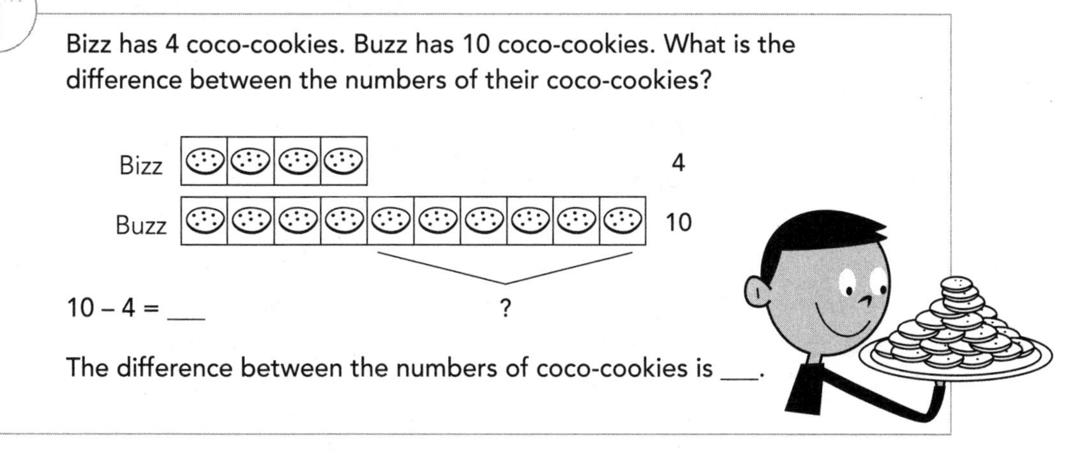

This comparison problem uses different wording in the question, specifically asking us to identify the difference between the quantities. The wording may be new to children, so discuss with them what is being asked.
Ask: *What is the question asking us to find?* (The difference between the number of Bizz and Buzz's coco-cookies)

Explain: *That's another way of asking how many more coco-cookies does Buzz have than Bizz. It's a comparison problem, just like the last problem.*

Point to the two rows of cookies on the board. Then click on the next page to display another version of the diagram that shows a square drawn around each cookie.

Say: *Now we see a square around each cookie. That makes it easier to count them. See how the squares are lined up, one above the other. That makes it easier for us to compare the two rows. Also, notice how Buzz's row is now on top. You don't always have to draw the squares in the order they appear in the problem. You can move them around if that makes it easier to compare them.*

Ask: *How many cookies does Buzz have?* (10) *How many cookies does Bizz have?* (4) *How can we find the difference?*

Guide children in counting and comparing the two quantities. Then invite them to share and demonstrate strategies for solving the problem. As in most comparison problems, drawing a line between the two quantities helps emphasize the difference between the two rows (see below).

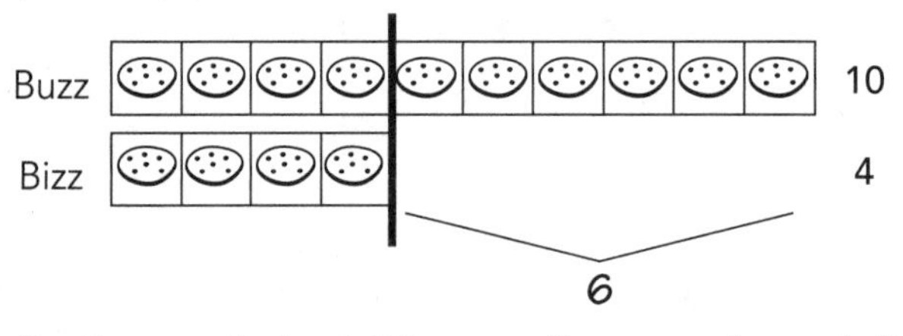

Drawing the line can help children easily see and count the difference by 1s or 2s. Have them complete the equation and the sentence by filling in the blanks.

Have children work on Problems #23 and 24 (page 67) in pairs. Remind them to draw unit squares around the pictures. Give children a few minutes to work. Then display each problem on the whiteboard. Invite children to share their strategies for solving the problems, interacting on the whiteboard whenever possible.

LESSON 7

Subtraction Using Unit Squares

Materials: student pages 68–69, pencils, projector, interactive whiteboard, markers

Preparation: Distribute copies of pages 68–69 and pencils to children. Go to www.scholastic.com/problemsolvedgr1 and click on Lesson 7. Set up your computer and projector to display the problems on the interactive whiteboard.

This lesson continues with basic subtraction, modeling both "take away" and "comparison" problems. But instead of pictures the problems will now use only squares to represent the quantities.

Display Problem #25 on the interactive whiteboard.

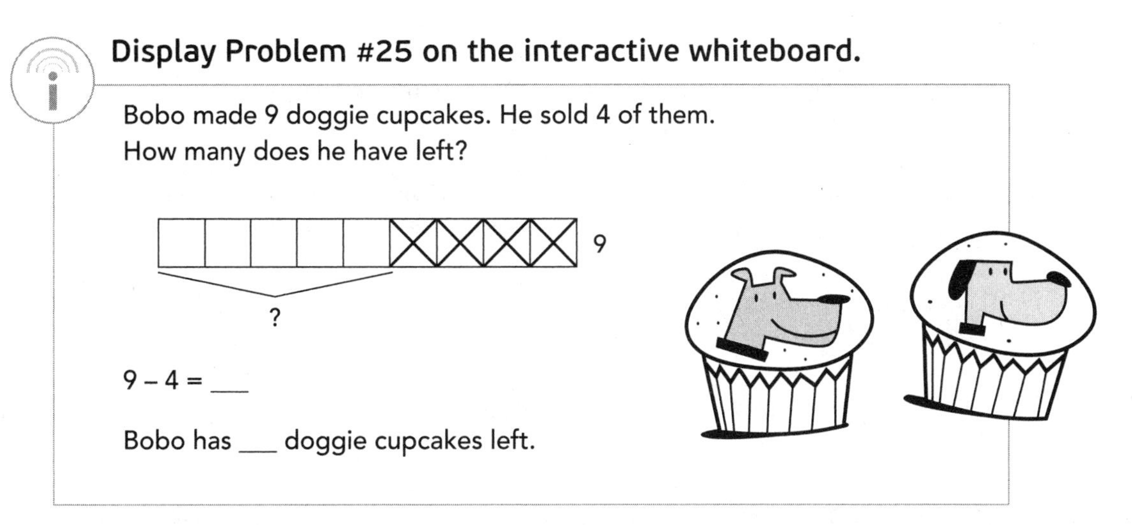

Read aloud the problem. Then have children underline the facts and circle the question on their papers.

Explain: *We are going to continue with subtraction in this lesson. But now we are going to take out the pictures and use just the squares. Drawing so many pictures can be hard and takes up a lot of time. But a square is easier to draw, and it can stand for different things in a problem.*

Ask: *In this problem, what does each square stand for?* (A doggie cupcake)

Say: *All of these squares are in one row. We are not comparing anything. We have some cupcakes and some are being sold or taken away. How can we show this?* (Cross out or shade in the cupcakes that were sold.)

Model crossing out or shading in the squares on the whiteboard. Then ask children to share their strategies for finding the answer. Some may decide to count the remainder by 1s or 2s. Others may count backwards 4 from 9. Encourage children to demonstrate their ideas on the board.

Display Problem #26 on the interactive whiteboard.

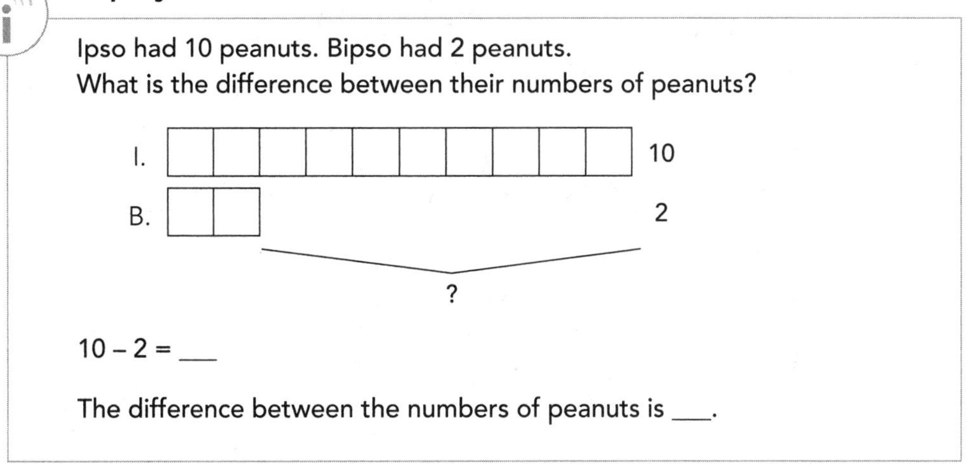

Read aloud the problem and ask children to identify the facts and the question. Point to the two rows of squares.

Say: *In the last problem, we had only one row of squares. Some things were taken away, and we crossed them out to find the difference. Here we have two rows of squares.*

Ask: *Why are the squares in two rows, one above the other?* (Because there are two different groups of peanuts, one for Ipso and one for Bipso.) *What are we doing with these two amounts?* (We are comparing them and finding the difference.)

Explain: *We are finding the difference between these two numbers. That means we are subtracting them, even though they are shown in two rows. We are not adding them up.*

Point out the labels and where the arrow and question mark are placed. Explain how these labels and markings help us clearly see the facts of the problem.

Guide children in comparing and counting the difference between the two quantities. Draw a line as before to highlight the comparison. The difference may be determined efficiently by counting by 2s or finding doubles, such as 3 + 3 or 4 + 4. Invite children to share their strategies on the board. Then have them complete the equation and sentence by filling in the blanks.

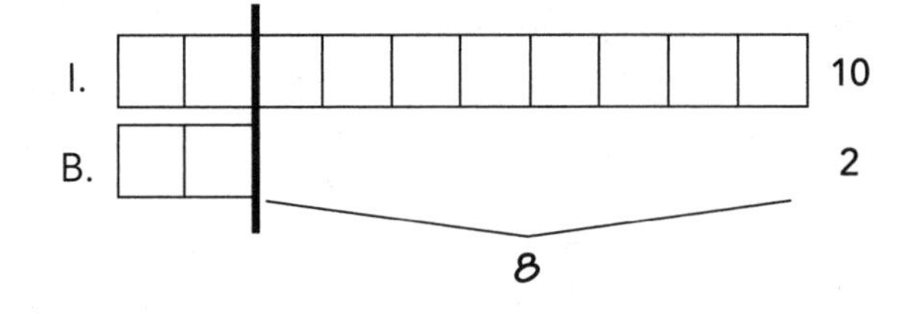

Have children work on Problems #27 and 28 (page 69) in pairs. Remind them to cross out squares as they subtract. Give them a few minutes to work. Then display each problem on the whiteboard. Invite children to share their strategies for solving the problems, interacting on the whiteboard whenever possible.

LESSON 8

Drawing Unit Squares to Subtract

Materials: student pages 70–71, pencils, projector, interactive whiteboard, markers

Preparation: Distribute copies of pages 70–71 and pencils to children. Go to www.scholastic.com/problemsolvedgr1 and click on Lesson 8. Set up your computer and projector to display the problems on the interactive whiteboard.

Children will draw their own unit squares to represent the quantities in the problems in this lesson. They will also label the squares with numbers, names, and an arrow with a question mark for the unknown quantity.

Display Problem #29 on the interactive whiteboard.

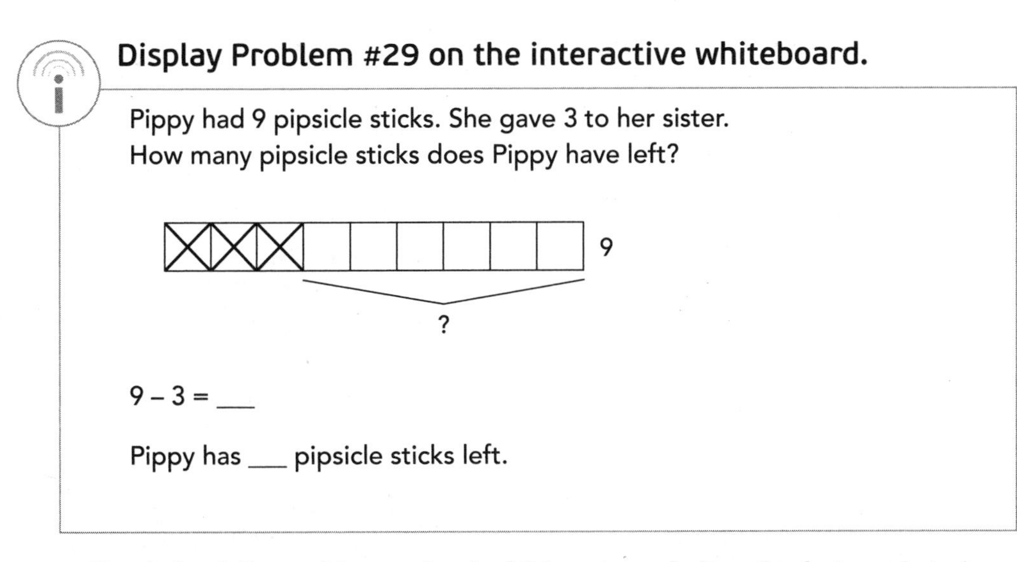

Read aloud the problem and ask children to underline the facts and circle the question on their papers.

Say: *In this lesson you will have to draw squares to show the things in the problem, like pipsicle sticks.*

Ask: *How many pipsicle sticks did Pippy have at the start of the problem?* (9) *How can we show this?* (Draw a square for each pipsicle stick.)

Call on a volunteer to draw 9 squares linked together on the board and label the quantity at the end of the row. Have children do the same on their papers. Remind them that the squares don't have to be perfect. They can use the gray lines on their paper to draw the squares neatly.

Say: *Pippy started with 9 pipsicle sticks. Then what happened?* (She gave 3 to her sister.) *How can we show this?* (Cross out or shade in 3 squares.)

Call on another volunteer to cross out 3 squares on the board. Have the rest of the class do the same on their papers.

Ask: *What are we trying to find out?* (How many pipsicle sticks are left)

Model how to draw an arrow and question mark to show where this unknown quantity is on the diagram. Have children follow on their papers.

Invite children to share their strategies for counting the remainder. They might count by 1s or 2s. Then have them fill in the blanks to complete the equation and sentence.

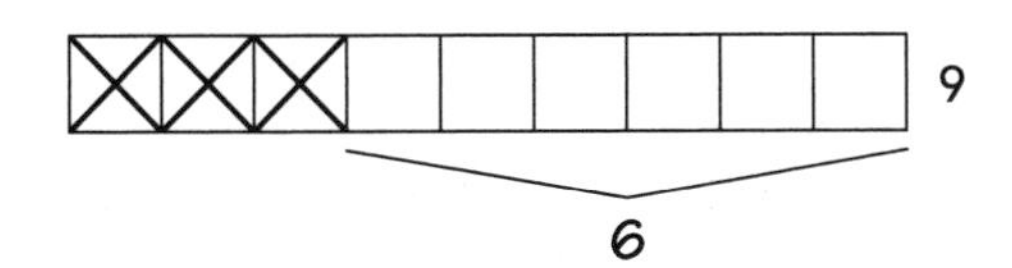

Display Problem #30 on the interactive whiteboard.

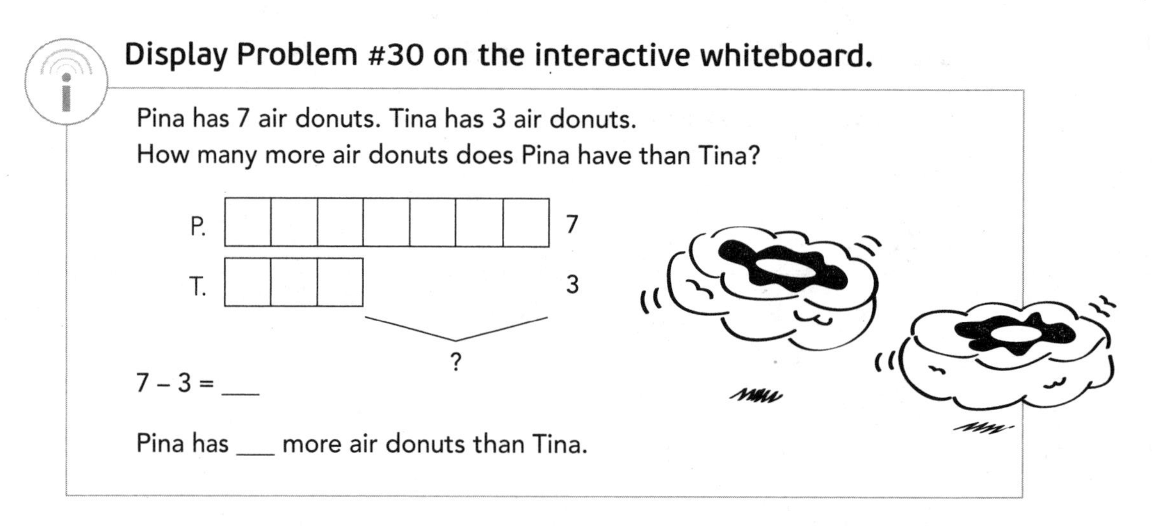

Read aloud the problem. Have children underline the facts and circle the question on their papers.

Say: *Here we have two amounts that we're comparing: Pina's air donuts and Tina's air donuts. How should we show the air donuts in this problem?* (Draw a row of squares for Pina's donuts and another row of squares underneath for Tina's donuts.) *Why do we need to show two* rows *of squares?* (Because we have two separate amounts that we are comparing.)

Have children draw the rows of squares on their paper, using the gray lines as a guide to draw the squares neatly. Model how to label the diagrams with abbreviations that show which row of squares belongs to whom, quantities at the end of each row, and an arrow and question mark to indicate the unknown.

Invite children to share their strategies for solving the problem. One effective strategy, for example, is to draw a line from the end of the 3 row up to the 7 row to show where the 3 in the first row matches up. Then children can count the difference between the two quantities. Have children fill in the blanks to complete the equation and sentence.

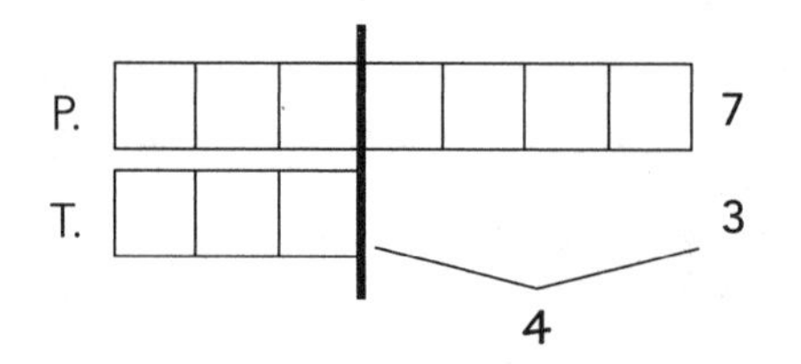

Have children work on Problems #31 and 32 (page 71) in pairs. Remind them to draw unit squares to represent the amounts in each problem and to label them. Give children a few minutes to work. Then display each problem on the whiteboard. Call on volunteers to draw and label the diagrams on the board. Then invite children to share their strategies for solving the problems, interacting on the whiteboard whenever possible.

Using Connected Squares With Addition and Subtraction (Within 20)

As the lessons move to addition and subtraction problems with larger numbers, children will use squares to make 10s or find doubles. Problems with three addends will be introduced in the addition lessons. Diagrams will be provided so children can focus on the strategies for solving the problems. However, they will still be expected to label the bars of connected squares properly, including drawing the arrow and question mark that indicate the unknown quantity.

Continue to follow the same lesson routine established earlier: read aloud the problems, go through the problem-solving process (identify and underline the facts and circle the question), and share and discuss solutions. Have children work in pairs as much as possible.

LESSON 9

Addition With Larger Numbers

Materials: student pages 72–73, pencils, projector, interactive whiteboard, markers
Preparation: Distribute copies of pages 72–73 and pencils to children. Go to www.scholastic.com/problemsolvedgr1 and click on Lesson 9. Set up your computer and projector to display the problems on the interactive whiteboard.

This lesson introduces a few strategies to help children tackle problems with larger numbers, such as making 10s or finding doubles. Remind children to label diagrams with numbers and words or abbreviations (when helpful), as well as an arrow and question mark for the unknown quantity.

Display Problem #33 on the interactive whiteboard.

Smedley had 6 Zorbo balls. He bought 7 more at the store.
How many Zorbo balls does he have now?

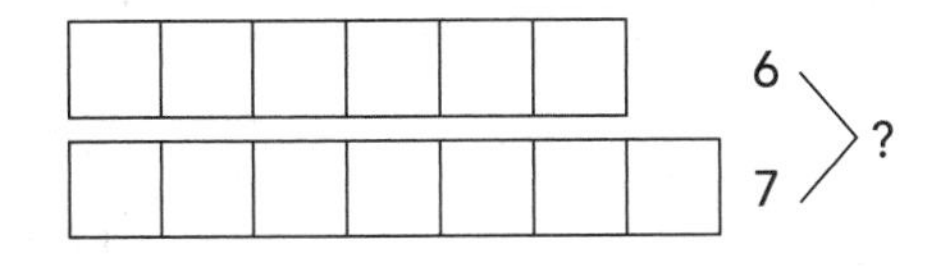

6 + 7 = ___

Smedley has ___ Zorbo balls now.

Read aloud the problem. Have children underline the facts and circle the question on their papers. Point out that the Zorbo balls in this problem are represented by two bars of connected squares. Guide children to label the diagram by writing in numbers for the quantities and drawing the arrow with a question mark for the unknown. Model this on the board and check children's work on their papers.

Have children look carefully at the way the two bars are set up and ask if they can think of any strategies for adding these two numbers. Share the diagram below on the board, which shows an outline around the top 5 and the bottom 5 squares.

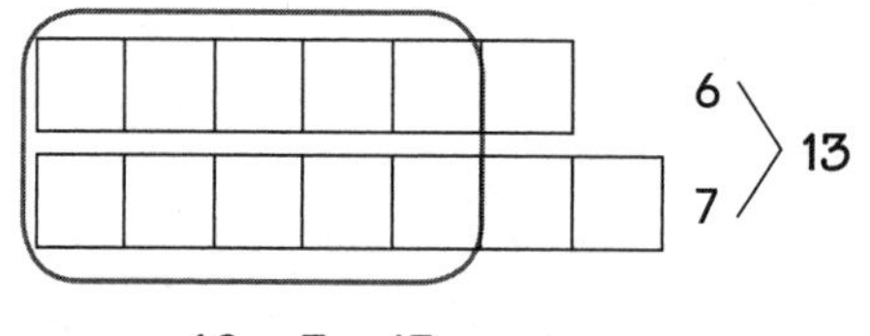

10 + 3 = 13

Explain: *Sometimes we can find easier problems within a bigger problem to help us. Can you see the 5 + 5 squares? Some of you already know your doubles facts, such as 5 + 5. This gives us 10 squares plus 3 more. We can also find smaller doubles or count by 2s.*

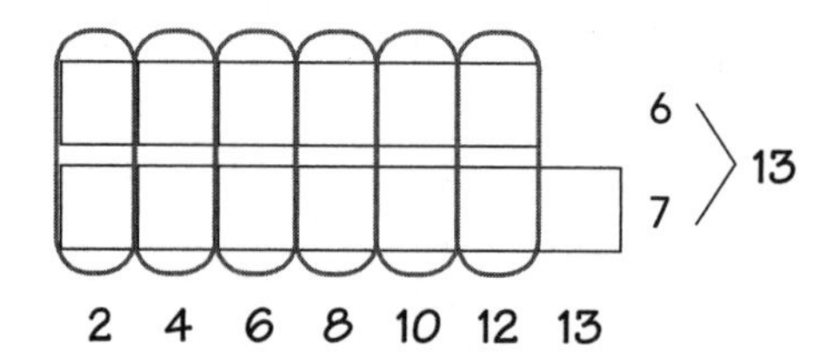

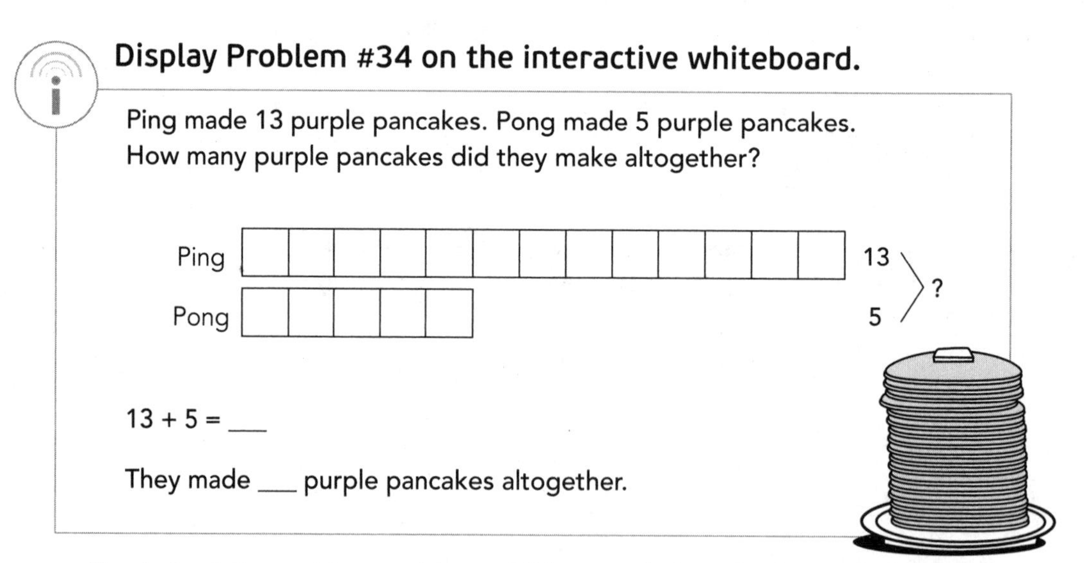

Read aloud the problem and have children underline the facts and circle the question on their papers.

Say: *We have two bars of connected squares, but no labels. Let's start by labeling them.* Guide children to label the diagram by writing the appropriate name and number for each bar and drawing the arrow and question mark to show the unknown. Then call on a volunteer to label the diagram on the board.
Explain: *Here we have one bar of 13 squares and one bar of 5 squares. In the last problem we found 5 + 5 squares. There's one here too.* Circle the squares as shown below, then invite children to count the remaining squares by 1s or 2s.

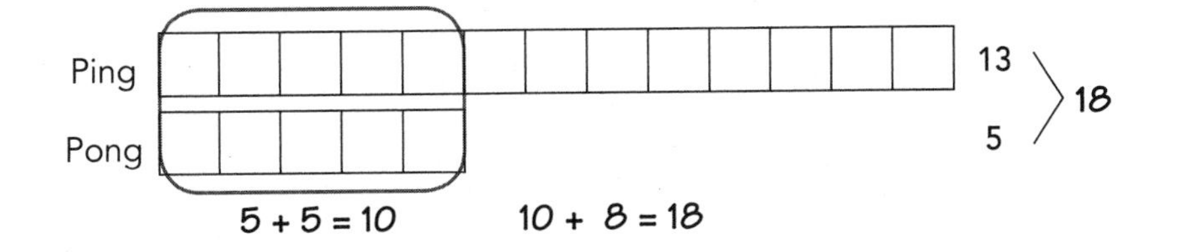

Say: *We can also use place value to help us with larger numbers, like 13. We can split this number into tens and ones. How many tens are in 13?* (1) *How many ones?* (3) Circle the 1 ten on the top row and label it *10*. Then label the remaining squares *3*, as shown below.

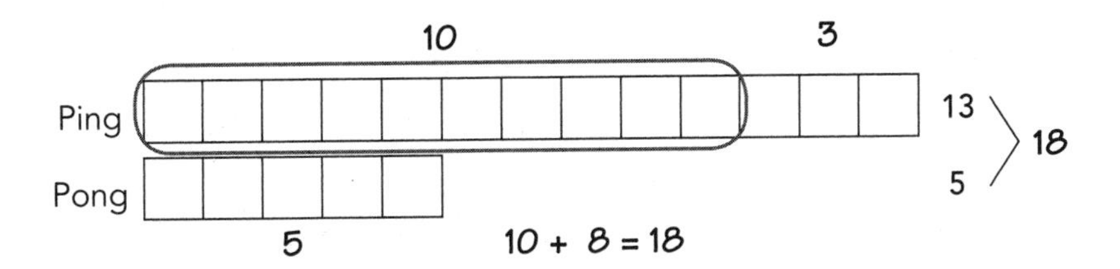

Ask: *Are there any tens in the second row?* (No) *How many ones are there?* (5)
Say: *So we can add the ones together: 3 + 5 = 8. We know we have 1 ten. Putting them together, we have 10 + 8 = 18. So the answer is 18. You can combine the squares any way you want to make easier additions.*

When we use an approach like this, it is helpful to show children that we can label a section of a bar with numbers to look like the diagram above. Encourage children to offer other possibilities and diagram these as well. Then have them try the strategies on their papers.

Have children work on Problems #35 and 36 (page 73) in pairs. Remind them to label the diagrams that represent each problem. Give children a few minutes to work. Then display each problem on the whiteboard. Call on volunteers to label the diagrams on the board. Then invite children to share their strategies for solving the problems, interacting on the whiteboard whenever possible.

LESSON 10

Addition With Three Addends

Materials: student pages 74–75, pencils, projector, interactive whiteboard, markers

Preparation: Distribute copies of pages 74–75 and pencils to children. Go to www.scholastic.com/problemsolvedgr1 and click on Lesson 10. Set up your computer and projector to display the problems on the interactive whiteboard.

The focus in this lesson is addition problems with three addends. Bars of connected squares are provided to acquaint children with the concept of representing three quantities.

Display Problem #37 on the interactive whiteboard.

Tom and Tim each has 5 bananas. Tip has 6 bananas.
How many bananas do the boys have altogether?

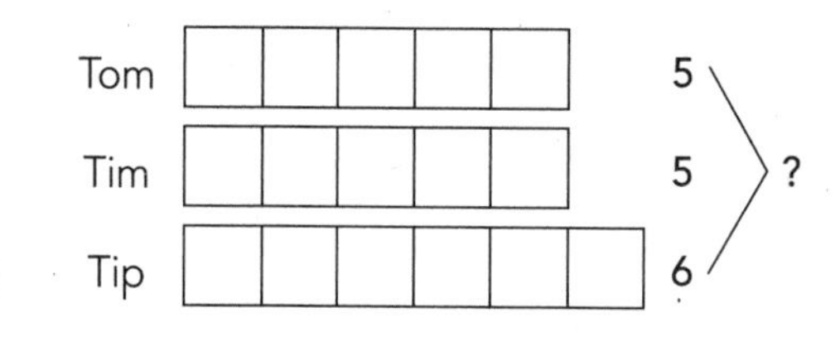

5 + 5 + 6 = ___

The boys have ___ bananas altogether.

Read aloud the problem. Have children identify the facts and question. Point out that this problem has three addends. Explain that this simply means we need three bars of connected squares instead of two.

Ask: *What do we know in this problem?* (Tom and Tim each has 5 bananas; Tip has 6.) *How many bars of squares do we need?* (3) *How many squares will be in the first bar?* (5) *What about the second bar?* (Also 5) *And what about the last bar?* (6)

Have children label the diagram on their papers, guiding them to write the appropriate name and number for each bar and draw an arrow and question mark to show the unknown. Then call on a volunteer to label the diagram on the board.

Invite children to share their strategies for combining these addends. Strategies may include counting by 5s to get to 15 and then adding 1 more, or adding two 5s to get 10 and then adding 6 more to get 16. Draw outlines around these squares to emphasize the strategies, as shown on the next page.

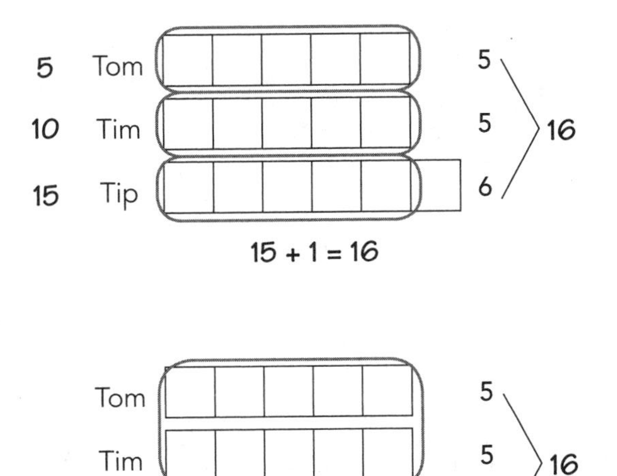

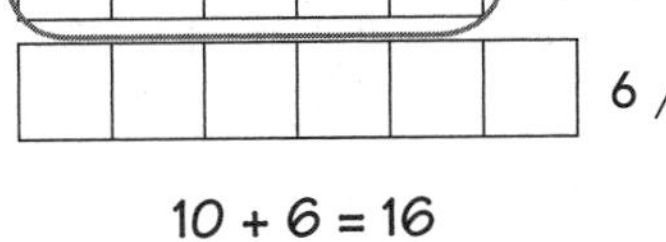

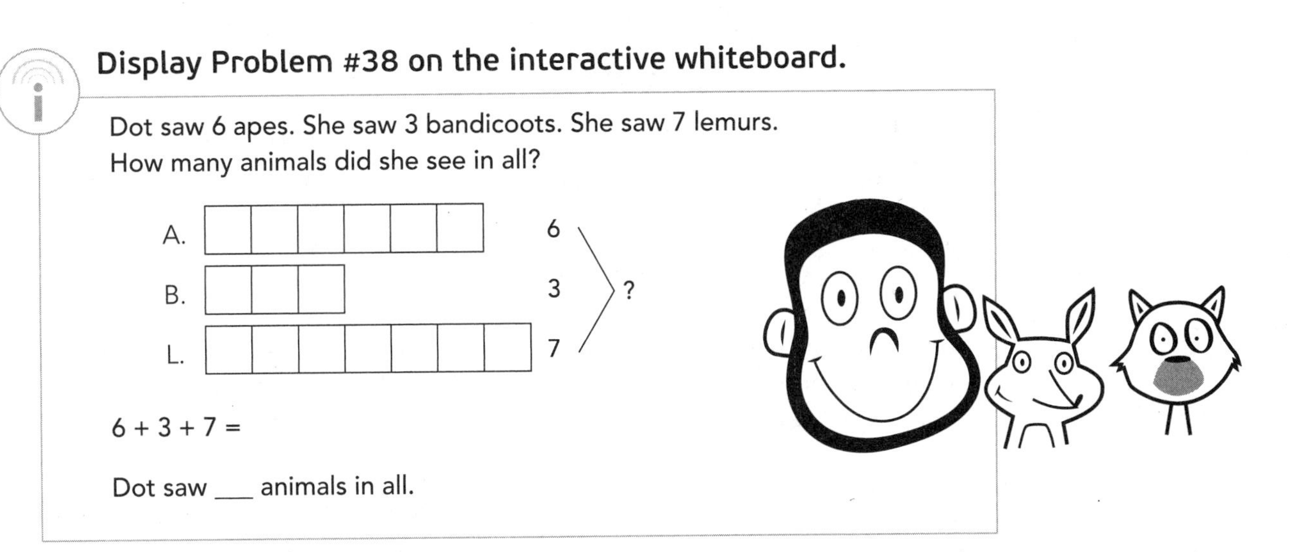

Read aloud the problem and have children identify the facts and question.
Say: *Here we have another problem with three addends. What are those addends?* (6, 3, and 7) *Let's label the diagrams to show the different animals. We'll put a letter in front of each bar to show which animal it represents. We'll also put a number at the end of each bar. Then we'll show the unknown with an arrow and question.* Call on volunteers to do these steps on the board.
Say: *We have three amounts to add. What are some ways to do this?*

Encourage children to share their strategies and draw outlines on the board to demonstrate their thinking. An easy strategy would be to make 10s; for example, 3 + 7.
Say: *We do not have to draw the bars in the same order as in the problem. We can move them around if that makes adding easier. For example, some people might like this arrangement better.* Share the diagram on page 32 on the board.

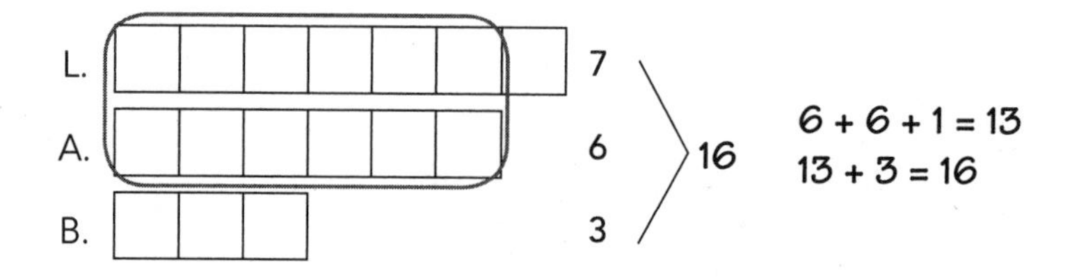

Reference Chart

To support children's understanding, you may want to create a reference chart for Bar Modeling to display somewhere in the classroom. On the chart consider including diagrams, vocabulary, and examples of different types of problems and strategies.

Explain: *Here we see the lemurs with 7 on top of the apes with 6. We can see that we are adding 6 + 6 + 1 = 13. Then all we have to do is add 3 more to make 16. Sometimes it's easier to see when similar quantities are closer to each other.*

Have children work on Problems #39 and 40 (page 75) in pairs. Remind them to label the diagrams that represent each problem. Give children a few minutes to work. Then display each problem on the whiteboard. Call on volunteers to label the diagrams on the board. Then invite children to share their strategies for solving the problems, interacting on the whiteboard whenever possible.

LESSON 11

Subtraction With Larger Numbers

Materials: student pages 76–77, pencils, projector, interactive whiteboard, markers

Preparation: Distribute copies of pages 76–77 and pencils to children. Go to www.scholastic.com/problemsolvedgr1 and click on Lesson 11. Set up your computer and projector to display the problems on the interactive whiteboard.

In this lesson children will work on subtraction, both "take away" and "comparison" problems, involving larger numbers. As in the previous two lessons, the diagrams will be provided and children will label them.

Display Problem #41 on the interactive whiteboard.

There were 15 books on the shelf. Bingo the bird took 5 of the books. How many books are left on the shelf?

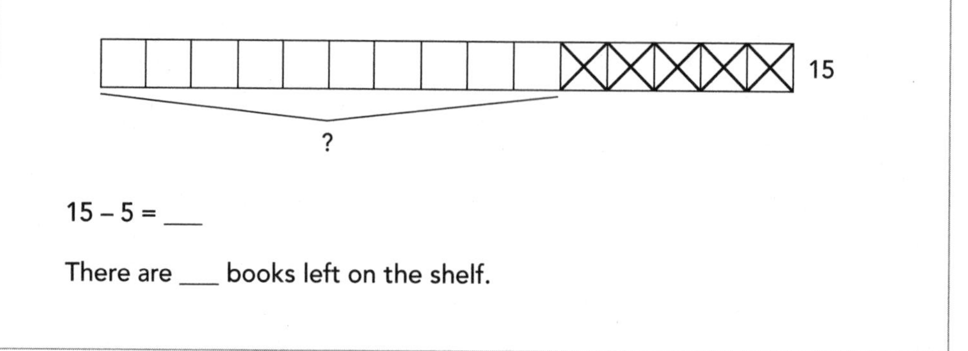

15 – 5 = ___

There are ___ books left on the shelf.

Read aloud the problem and have children underline the facts and circle the question on their papers.

Say: *In today's lesson we'll see subtraction problems with larger numbers. Let's start with this problem on the board. We see a bar of connected squares.*
Ask: *What do these 15 squares show?* (15 books) *What happened to these books in the problem?* (Bingo took 5 of the books.)
Say: *Let's label this diagram and draw an arrow and question mark to show the unknown.* Model how to label the bar with a number and show the unknown with an arrow and a question mark. Have children do the same on their papers. Point out that since the problem is only about one set of books, we don't need to label the diagram "books."
Ask: *How can we solve this problem using our diagram?* (Cross out or shade in 5 squares, then count the remaining squares.)

Call on a volunteer to demonstrate this strategy on the board. Then invite children to share ideas about how to find the difference. They can count backwards from 15 or count the remaining squares that are not crossed out.

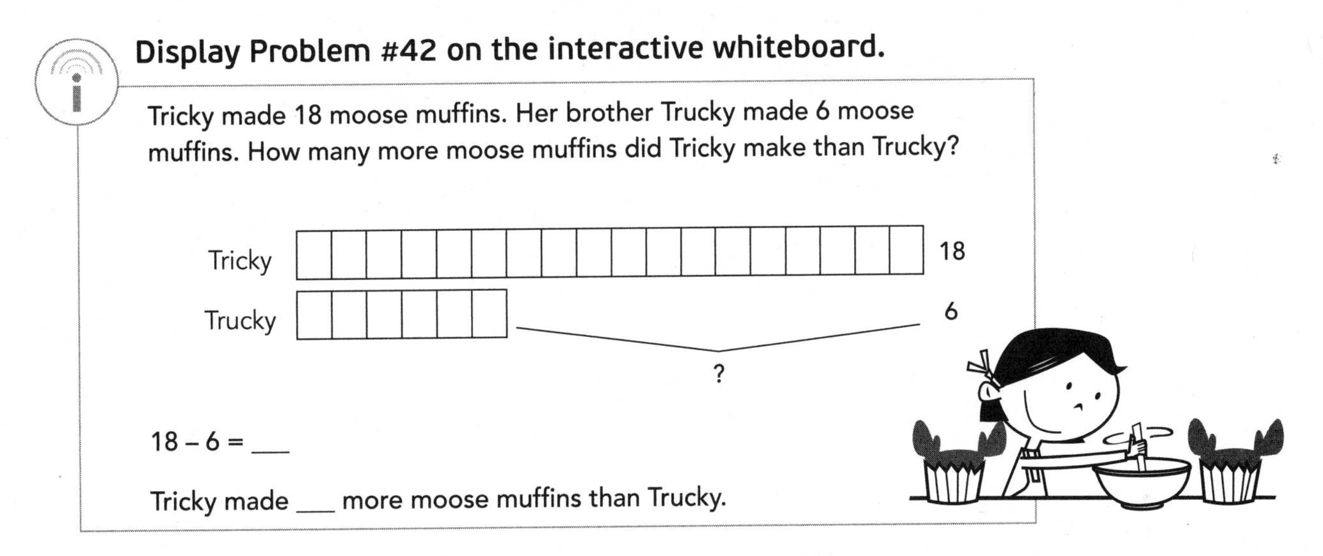

Ask: *What kind of problem is this?* (A comparison problem) *What are we comparing?* (How many more moose muffins Tricky made compared to Trucky)
Say: *Remember when we have a comparison problem, we want to put one amount above the other. This makes it easier to compare the two amounts. You can see the two bars of connected squares on your paper. Let's label them with names, numbers, and an arrow and question mark to show the unknown.* Model how to label the bars on the board.

Invite children to share their strategies for solving the problem and showing them on the board. Possibilities include drawing a line at the 6 in each row and counting the difference by 2s.

Have children work on Problems #43 and 44 (page 77) in pairs. Remind them to label the diagrams that represent each problem. Give children a few minutes to work. Then display each problem on the whiteboard. Call on volunteers to label the diagrams on the board. Then invite children to share their strategies for solving the problems, interacting on the whiteboard whenever possible.

More Subtraction With Larger Numbers

Materials: student pages 78–79, pencils, projector, interactive whiteboard, markers

Preparation: Distribute copies of pages 78–79 and pencils to children. Go to www.scholastic.com/problemsolvedgr1 and click on Lesson 12. Set up your computer and projector to display the problems on the interactive whiteboard.

As children work with larger numbers, using connected squares will help support them as they try out various strategies. Diagrams without labels will be provided.

Display Problem #45 on the interactive whiteboard.

There were 20 boats at the dock. 10 of the boats sailed to Boo Boo Island. How many boats are still at the dock?

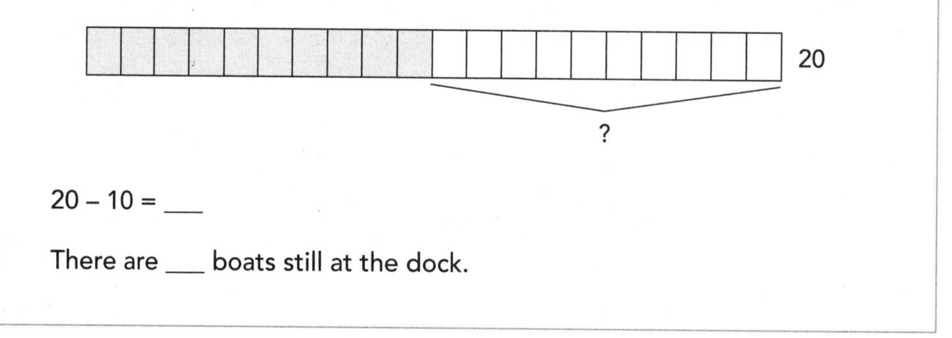

20 – 10 = ___

There are ___ boats still at the dock.

Read aloud the problem. Ask children to underline the facts and circle the question on their papers.

Say: *This problem comes with a bar of connected squares, but we still need to label it. Let's label the bar* 20 *to show the 20 boats.* Model labeling the diagram on the board. Have children follow along on their papers.

Ask: *Now, what happened to some of the boats in the problem?* (They sailed away.) *How many sailed away?* (10)

Say: *To show the boats that sailed away, we can cross out or shade in 10 squares in our bar. Then we'll label this part* 10. Have children do the same on their papers.

Ask: *What does this other part of the bar show?* (The boats that have not sailed away) Draw an arrow and question mark under this section, and have children do the same.

Ask: *How many boats do you think are still at the dock?*

Encourage children to share their strategies and answers. Some children may know the 10 + 10 double and use that to determine the unknown. Another strategy is to section the squares in 2s and count up, as shown on the following page.

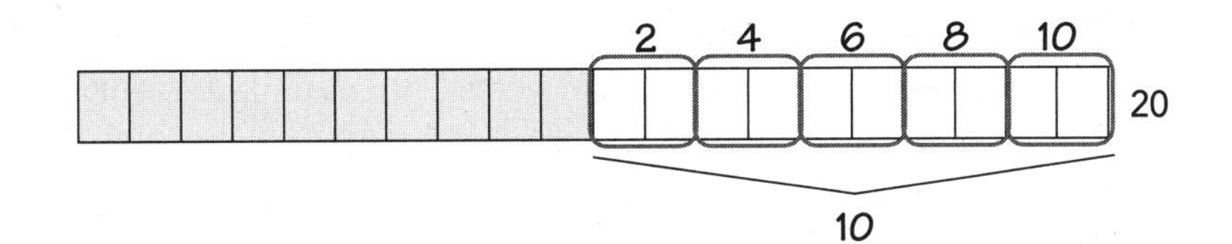

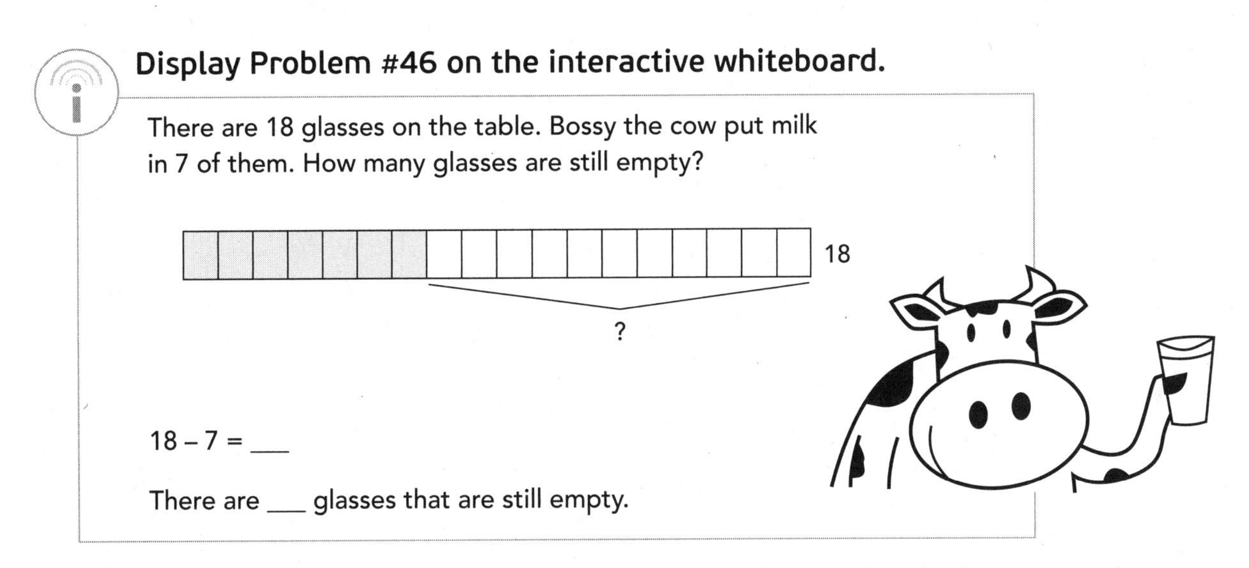

Display Problem #46 on the interactive whiteboard.

There are 18 glasses on the table. Bossy the cow put milk in 7 of them. How many glasses are still empty?

18 – 7 = ___

There are ___ glasses that are still empty.

Read aloud the problem. Have children underline the facts and circle the question on their papers.

Say: *We have another problem here with larger numbers and a diagram. Let's label the diagram to show the situation.* On the board label the end of the row *18*. Have children do the same on their papers.

Ask: *What happened next in the problem?* (Bossy filled 7 glasses with milk.)

Say: *I want to show the glasses that got milk. So I am going to cross out or shade in 7 squares.* As you model this on the board, have children do the same on their papers. Remind them that it doesn't matter from which side they start crossing out squares.

Say: *Now, what does this empty part of the bar stand for?* (The glasses without milk) *Let's mark that with an arrow and a question mark.*

Encourage children to work in pairs to solve the problem. Call on volunteers to share their ideas on the board. One possibility is to mark 3 more squares, which added to 7 is 10, as shown below. The remaining portion of the bar must be 8, so 3 + 8 = 11. So there are 11 glasses that are still empty.

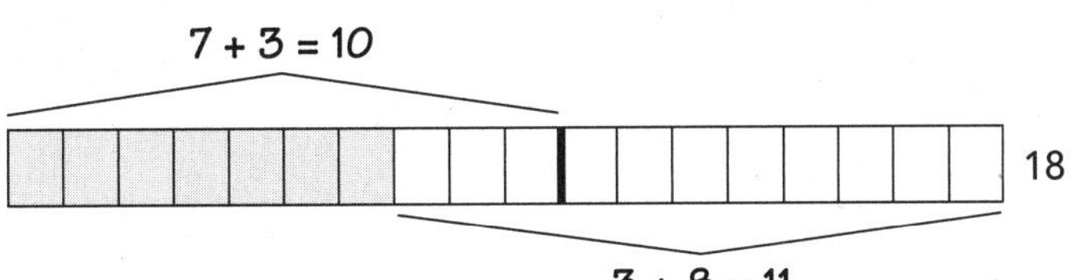

This "Finding Ten" strategy in subtraction is a new sophisticated idea for children and may need a good deal of practice for them to gain a solid understanding.

Have children work on Problems #47 and 48 (page 79) in pairs. Remind them to label the diagrams that represent each problem. Give children a few minutes to work. Then display each problem on the whiteboard. Call on volunteers to label the diagrams on the board. Then invite children to share their strategies for solving the problems, interacting on the whiteboard whenever possible.

Using Connected Squares With Measurement Problems

In this chapter children will solve measurement word problems using both addition and subtraction. We will rely on the connected squares diagram as a strategy, with children drawing some of the diagrams themselves. From this point on, children will also write their own equations to match their diagrams.

Follow the same lesson routine established earlier: read aloud the problems together, go through the problem-solving process (identify and underline the facts and circle the question), and share and discuss solutions. As much as possible, have children work in pairs.

LESSON 13

Measurement Addition Problems

Materials: student pages 80–81, pencils, projector, interactive whiteboard, markers

Preparation: Distribute copies of pages 80–81 and pencils to children. Go to www.scholastic.com/problemsolvedgr1 and click on Lesson 13. Set up your computer and projector to display the problems on the interactive whiteboard.

Measurement is the context for the problems in this lesson, with a focus on addition. Remind children to properly label the diagrams with names and numbers, as well as the arrow and question mark for the unknown quantity. In this lesson children will also need to write equations to match their work for the first time.

Display Problem #49 on the interactive whiteboard.

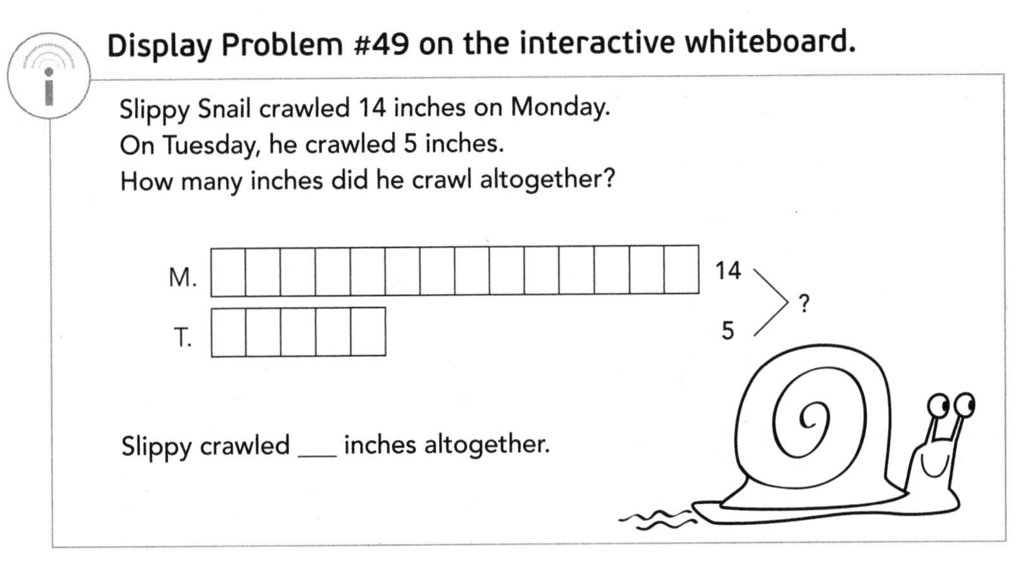

Read aloud the problem. Have children underline the facts and circle the question on their papers.

Say: *Today's problem involves length—how long or how far something is. What unit of measure are they talking about in this problem?* (Inches) *We can use our bars to show length.*
Ask: *Which bar of squares is longer?* (Monday) *Label the bars with letters and numbers and draw an arrow and question mark for the unknown.*

Call on volunteers to label the diagram on the board.

Have children work in pairs to solve the problem. Point out that there is no equation on their paper. Have them write an equation that matches their work. Then invite volunteers to share their strategies and solutions on the board. Some children may see the relationship shown below.

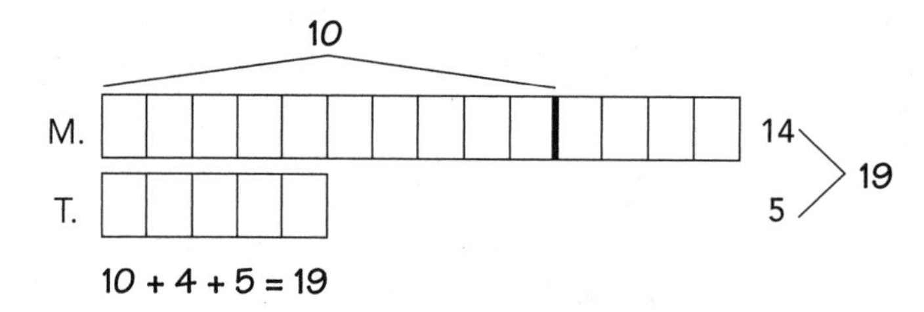

In this strategy children can split the top row into tens and ones, then add the ones together, plus 10: 10 + 4 + 5 = 19. Children can also make groups of 5s and count three 5s plus 4 to get to 19. Encourage this kind of part-whole flexible number thinking. If children do not come up with it on their own, model this strategy so they can begin to see it as a possibility.

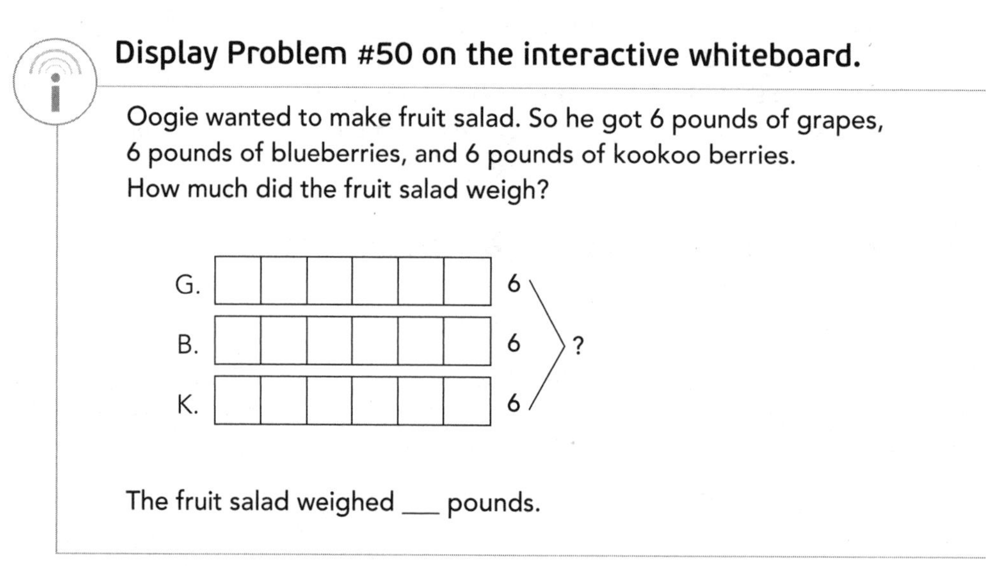

Read aloud the problem. Have children underline the facts and circle the question on their papers. Point out that this problem does not include a diagram.

Ask: *How many amounts do we have in this problem?* (3) *So how many bars of connected squares do we need to draw?* (3) *How many squares in each bar?* (6)

Have children draw the diagram to show the three amounts and label it as usual. Call on volunteers to share their diagrams on the board. Then have children work in pairs to solve the problem. Remind them to write an equation to match their work. Invite children to share their strategies and solutions on the board.

Here is a strategy to share with children, if no one comes up with it. Share the diagram below on the board and demonstrate how we could find groups of 5 to make 10 or even 15. Then simply add on the leftover 3.

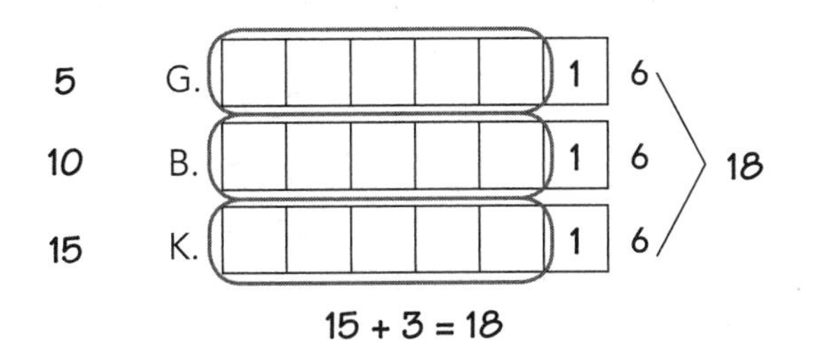

Have children work on Problems #51 and 52 (page 81) in pairs. Remind them to draw and label diagrams to represent the amounts in each problem. Have them write the equations too. Give children a few minutes to work. Then display each problem on the whiteboard. Call on volunteers to draw the diagrams and write the equations on the board. Then invite children to share their strategies for solving the problems, interacting on the whiteboard whenever possible.

LESSON 14

Measurement Subtraction Problems

Materials: student pages 82–83, pencils, projector, interactive whiteboard, markers

Preparation: Distribute copies of pages 82–83 and pencils to children. Go to www.scholastic.com/problemsolvedgr1 and click on Lesson 14. Set up your computer and projector to display the problems on the interactive whiteboard.

This lesson continues with measurement problems, but this time with a focus on subtraction. As in the previous lesson, children will draw some of the diagrams themselves to represent the quantities. Remind them to label the diagrams and draw an arrow and question mark to show the unknown.

Display Problem #53 on the interactive whiteboard.

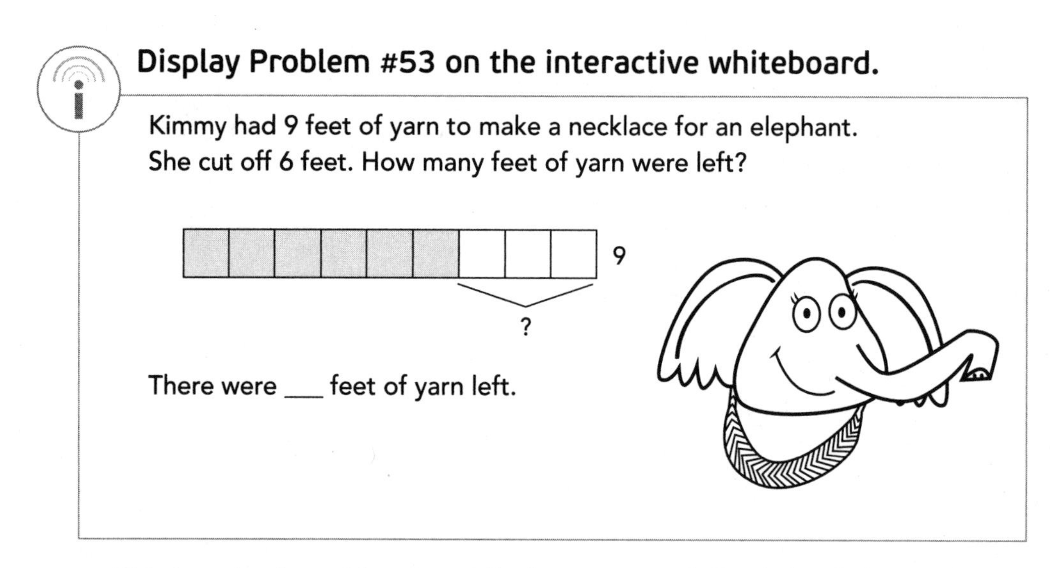

Kimmy had 9 feet of yarn to make a necklace for an elephant.
She cut off 6 feet. How many feet of yarn were left?

There were ___ feet of yarn left.

This is a simple problem to get the lesson started. Read aloud the problem and have children underline the facts and circle the question on their papers. **Ask:** *What kind of subtraction problem is this—take away or comparison?* (Take away) *So what will our diagram look like?* (One bar of 9 connected squares)

Guide children to draw the diagram on their papers to show the quantity being subtracted from. They should label the bar *9*. To show the 6 being taken away, have children cross out or shade in 6 squares.

Have children work in pairs to determine the value of the remaining section. This is most often determined by counting up from 6 to reach 9, with the result being 3. Make sure children write the equation to match this problem.

Display Problem #54 on the interactive whiteboard.

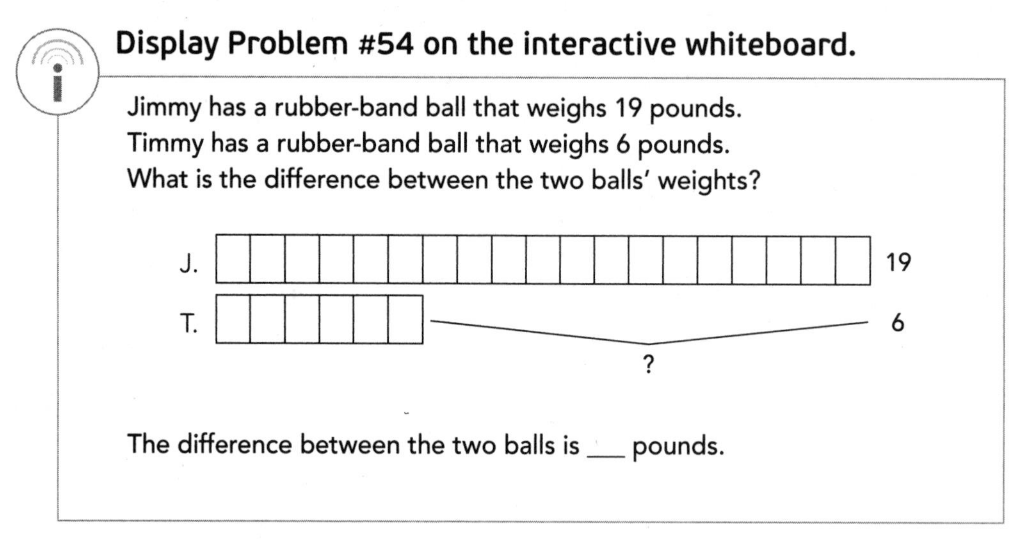

Jimmy has a rubber-band ball that weighs 19 pounds.
Timmy has a rubber-band ball that weighs 6 pounds.
What is the difference between the two balls' weights?

The difference between the two balls is ___ pounds.

Read aloud the problem and have children identify the facts and question. **Ask:** *What kind of subtraction problem is this—take away or comparison?* (Comparison)

Say: *In this problem we are looking at two different things being compared. What are those two things?* (Jimmy's rubber-band ball and Timmy's rubber-band ball) *Whose bar is longer?* (Jimmy's)

Have children label the diagram on their papers with names or initials and numbers, and draw an arrow and question mark to show the unknown. Remind them to write a matching equation as well. Then have them work in pairs to solve the problem and share solutions with the class.

Here is one strategy to share: Draw a line showing where the 6 in Jimmy's row matches up with the 6 in Timmy's row. Whatever is left over in Jimmy's row is the difference. To determine exactly what that would be, first split the remaining piece into a 4, which added to the 6 would be 10. The other remaining piece would have to be 9, because 6 + 4 + 9 = 19. This lets us see that the 4 + 9, or 13, is the difference (see below).

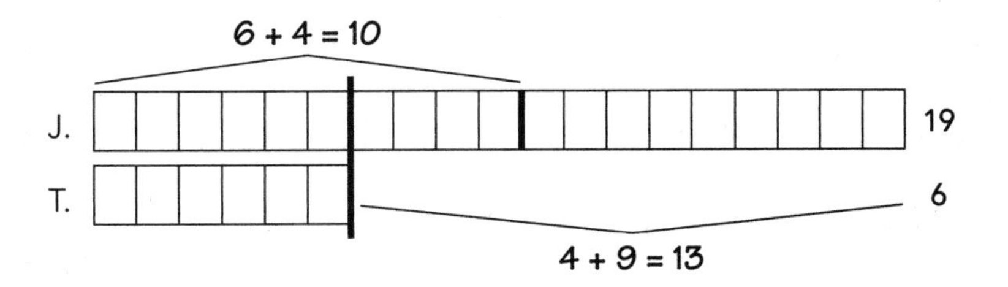

Have children work on Problems #55 and 56 (page 83) in pairs. Remind them to draw and label diagrams to represent the amounts in each problem. Have them write the equations too. Give children a few minutes to work. Then display each problem on the whiteboard. Call on volunteers to draw the diagrams and write the equations on the board. Then invite children to share their strategies for solving the problems, interacting on the whiteboard whenever possible.

CHAPTER 5

Using Connected Squares With Money Problems

The focus in this chapter is money, and children will continue to apply what they've learned about Bar Modeling to solve addition and subtraction problems.

Follow the same lesson routine established earlier: read aloud problems together, go through the problem-solving process (identify and underline the facts and circle the question), and share and discuss solutions. As much as possible, have children work in pairs.

LESSON 15

Money Addition Problems

Materials: student pages 84–85, pencils, projector, interactive whiteboard, markers

Preparation: Distribute copies of pages 84–85 and pencils to children. Go to www.scholastic.com/problemsolvedgr1 and click on Lesson 15. Set up your computer and projector to display the problems on the interactive whiteboard.

Even though the context of the problems in this lesson is money, the strategy should be familiar to children. They will draw bars of connected squares and write equations to help them solve the problems.

Display Problem #57 on the interactive whiteboard.

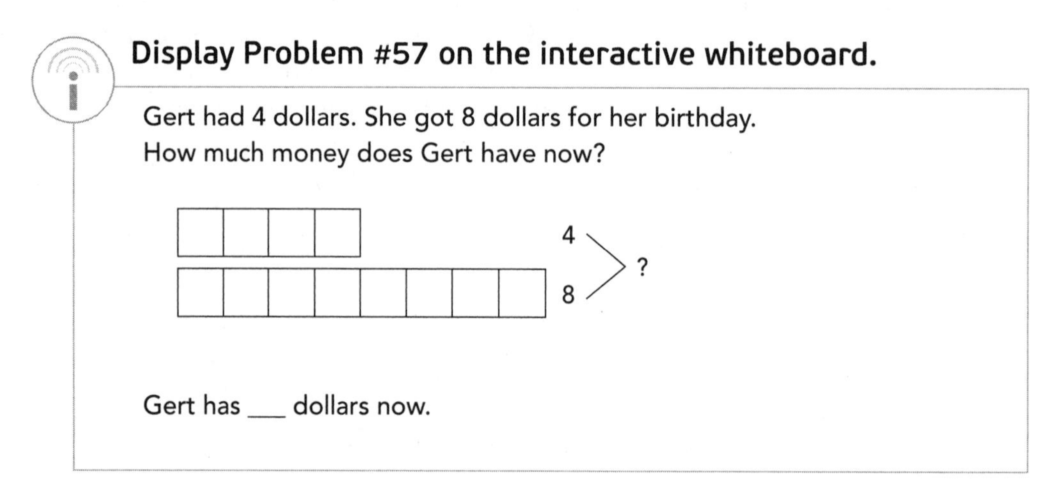

Read aloud the problem and ask children to identify the facts and question. **Say:** *This is like a regular addition problem we've done before. On your paper draw two bars of connected squares to show each amount, one above the other. Label the bars with numbers. Then draw an arrow and question mark to show the unknown. Then write an equation to match the problem.*

Call on volunteers to draw the diagram and write the equation on the board. Have children work in pairs to solve the problem, and then invite them to share their strategies and solutions on the board.

Remind the class that it is almost always helpful to look for a way to make 10. For example, we could split the 4 bar to show 2 + 2. We can then make 10 by adding 2 + 8, which leaves us with 2 extra. Together, that makes 12 (see below).

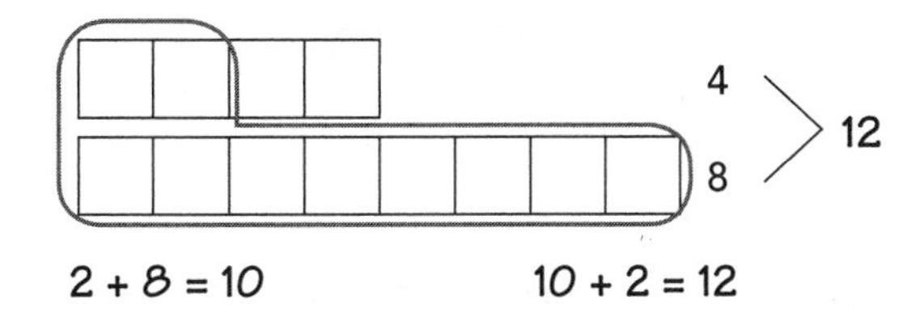

Display Problem #58 on the interactive whiteboard.

Malik found 4 cents in his sock. He found 4 more cents in his shoe and 5 cents in his ear. How much money did Malik find altogether?

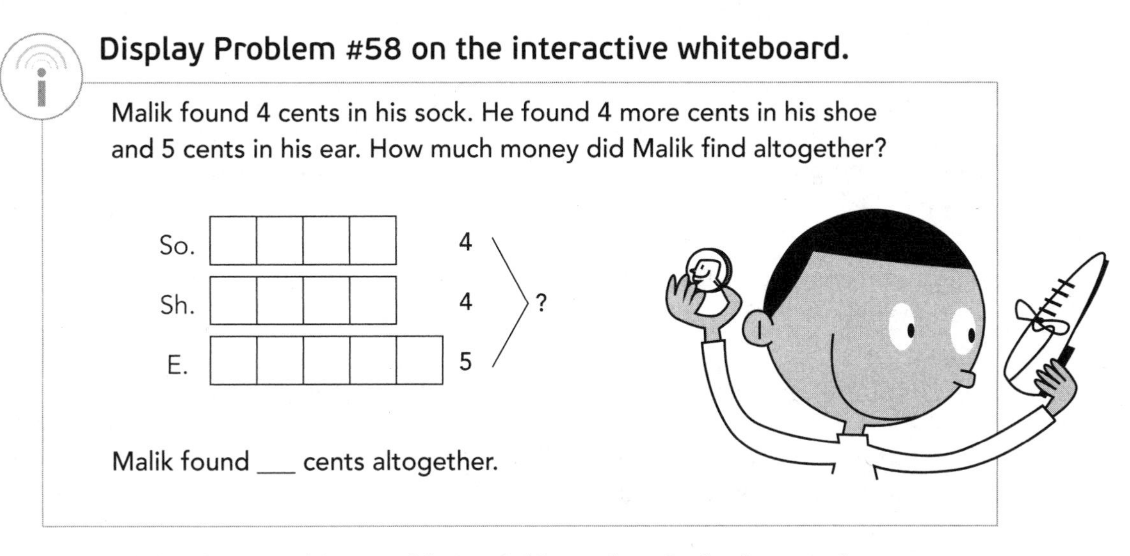

Malik found ___ cents altogether.

Read aloud the problem and have children identify the facts and question. **Ask:** *How many addends do we have in this problem?* (3) *What are they?* (4, 4, and 5) *So how many bars of connected squares do we need to draw for this problem?* (3)

Have children draw the diagram on their paper. Then have them label it and write a matching equation. Call on volunteers to share their diagrams and equations on the board.

Then ask children to work in pairs to solve the problem. Invite children to share their strategies and solutions on the board.

Again, remind the class to look for ways to make 10. In this problem, we could circle the two 4s to make an 8. Then we could split the 5 bar into 2 + 3, as shown below. This would leave 3 extra, giving us 10 + 3 = 13.

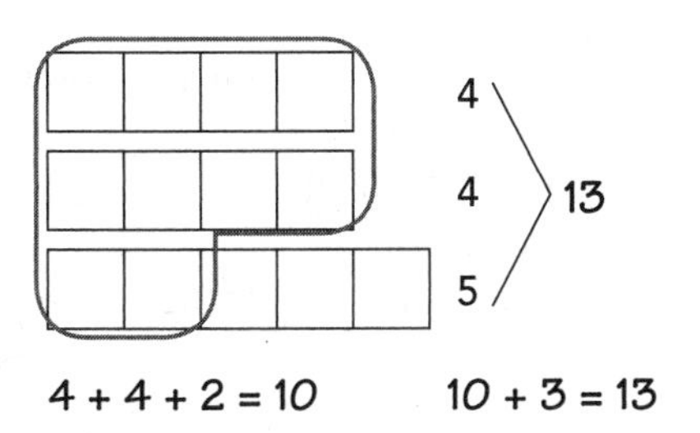

Have children work on Problems #59 and 60 (page 85) in pairs. Remind them to draw and label diagrams to represent the amounts in each problem. Have them write the equations too. Give children a few minutes to work. Then display each problem on the whiteboard. Call on volunteers to draw and label the diagrams and write the equations on the board. Then invite children to share their strategies for solving the problems, interacting on the whiteboard whenever possible.

LESSON 16

Money Subtraction Problems

Materials: student pages 86–87, pencils, projector, interactive whiteboard, markers

Preparation: Distribute copies of pages 86–87 and pencils to children. Go to www.scholastic.com/problemsolvedgr1 and click on Lesson 16. Set up your computer and projector to display the problems on the interactive whiteboard.

Children get to practice subtraction using Bar Modeling, all within the context of money. As in the previous lesson, children will draw their own diagrams and write their own equations to match the problem.

Display Problem #61 on the interactive whiteboard.

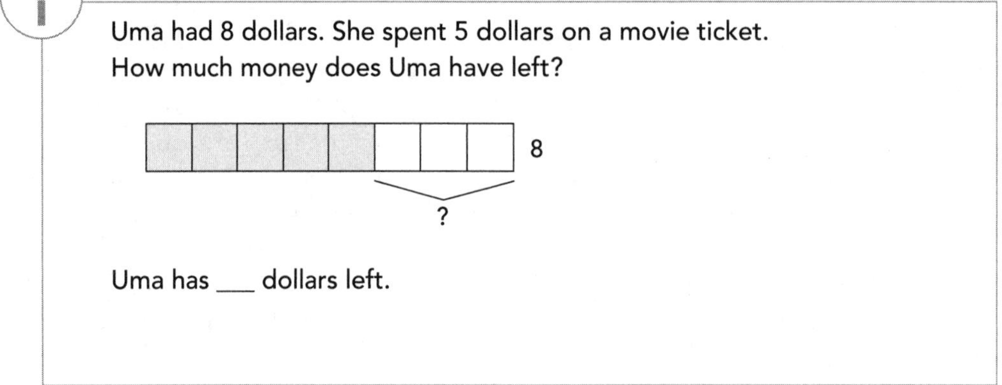

Read aloud the problem and have children identify the facts and question.

Ask: *What kind of subtraction problem do we have here—take away or comparison?* (Take away) *So how can we show this problem?* (Draw a bar of 8 connected squares.)

Say: *We need only one bar because we are starting with one amount, then taking away from that. Draw a diagram on your paper and label it.*

Ask: *How do we show the amount being taken away?* (Cross out or shade in 5 squares.)

Have children write an equation to match the problem. Then call on volunteers to share their diagrams and equations on the board.

To solve this, many children will count the remaining 3 squares that are not crossed out, or count down 5 from 8.

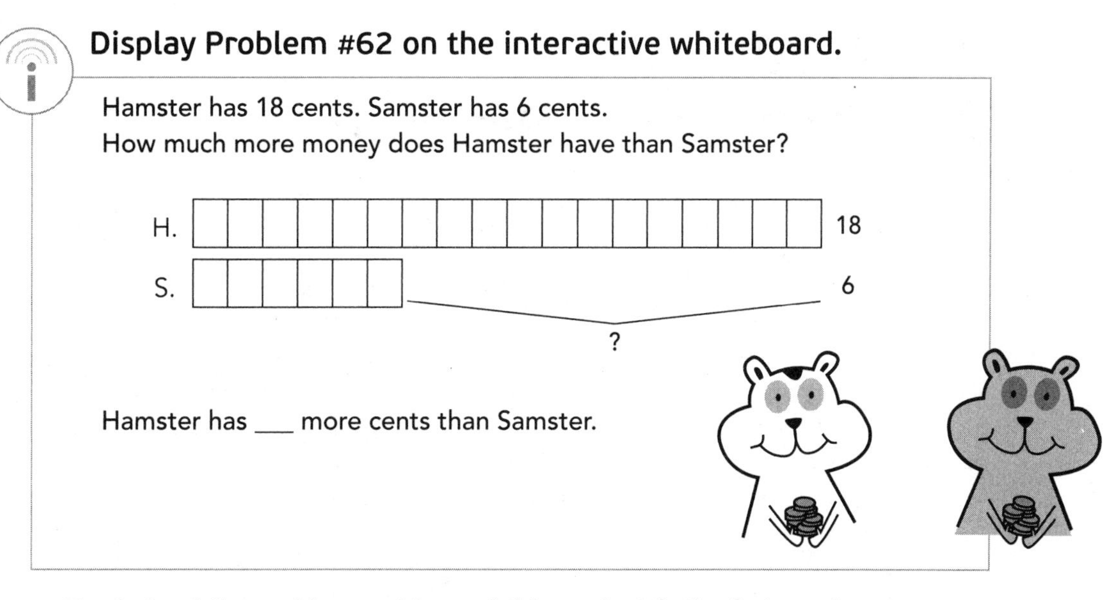

Read aloud the problem and have children identify the facts and question.
Say: *In this problem, we see two amounts are being compared. We want to find out how much more money Hamster has than Samster. On your paper you will see the diagram has already been drawn. You just need to label it.*
Ask: *Whose bar is longer?* (Hamster's)

Have children label the diagram with names or initials and numbers. Guide them where to draw the arrow and question mark, then have them write a matching equation. Call on volunteers to share their diagrams and equations on the board.

One strategy is to draw a line through the top bar that matches up with the 6 on the lower bar. This way we can see how much of the bars are equal. The remainder is the amount that is more than Samster's. Guide children to make 10 to find the difference (see below) or to count down 6 from 18.

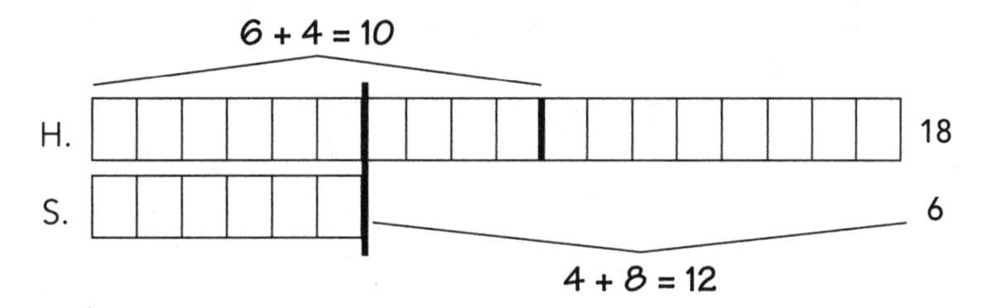

Have children work on Problems #63 and 64 (page 87) in pairs. Remind them to draw and label diagrams to represent the amounts in each problem. Have them write the equations too. Give children a few minutes to work. Then display each problem on the whiteboard. Call on volunteers to draw the diagrams and write the equations on the board. Then invite children to share their strategies for solving the problems, interacting on the whiteboard whenever possible.

CHAPTER 6

Bar Modeling With Multistep Problems

In this chapter children will combine everything they've learned about Bar Modeling so far to solve multistep problems involving addition and subtraction. Some problems may call for more than one bar model diagram, while others can be solved using only one diagram for all operations.

The lesson routine we've established earlier will be as valuable as ever: read aloud the problems together, go through the problem-solving process to identify the facts and question, and share and discuss solutions. Have children work in pairs as much as possible, then invite volunteers to share their diagrams, equations, and strategies on the board.

LESSON 17

Mixed Multistep Problems

Connected Squares vs. Continuous Bars

By now, drawing connected squares to represent the amounts in a problem should be second nature to children. If you feel your students are up to it, you may want to show them that they can erase the lines between the connected squares to show a continuous bar. As quantities get larger, it becomes harder and time-consuming to represent them using connected squares. A simple continuous bar or rectangle with appropriate labels can do the job.

Materials: student pages 88–89, pencils, projector, interactive whiteboard, markers

Preparation: Distribute copies of pages 88–89 and pencils to children. Go to www.scholastic.com/problemsolvedgr1 and click on Lesson 17. Set up your computer and projector to display the problems on the interactive whiteboard.

In this lesson children will encounter two-step word problems that use both addition and subtraction. Keeping track of all the facts becomes more challenging in these types of problems, but this is where Bar Modeling becomes most helpful.

Display Problem #65 on the interactive whiteboard.

Nelly Numbers had 4 math books. She bought 3 more. She gave 5 books to her sister. How many math books does Nelly have left?

Nelly has __ books left.

Read aloud the problem. Have children underline the facts and circle the question on their papers.

Say: *In this lesson we are going to start solving some problems that are a little more complex. We may need to add and subtract in the same problem. Because of that we might need to use more than one diagram. Let's start with what we know. We know that Nelly had 4 math books. Then she bought 3 more. How do we show that with our bar model?*

Model on the board how to draw and label two bars of connected squares that show this part of the problem. Write a matching equation as well (see Step 1 below). Have children do the same on their papers. Ask children to work in pairs to come up with the solution. When they determine that the result is 7, guide children into creating a diagram to show the second part of the problem. **Ask:** *What's the second part of the problem?* (She gave 5 books to her sister.) *And what is the question asking us to find?* (How many math books does Nelly have left?) *What operation do we need to do to find the answer?* (Subtraction) *How do we show this with our bar model?*

Guide children to understand that they need to draw a second diagram that shows 7 – 5. The result should look something like Step 2 below.

STEP 1

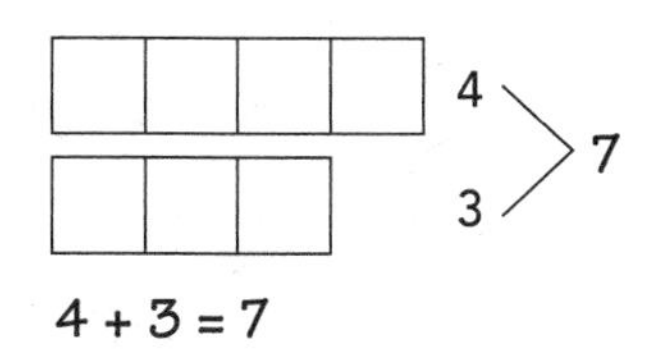

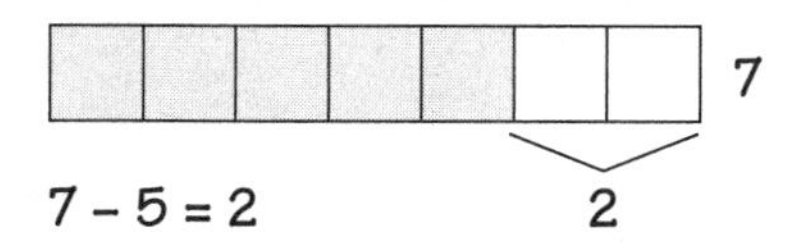

Tell children that it is also possible to use one diagram to complete all operations—addition and subtraction—as shown below. Starting with our first diagram (Step 1), cross out or shade in the 5 books Nelly gave to her sister. We can then see that there are 2 left. Share this diagram on the board.

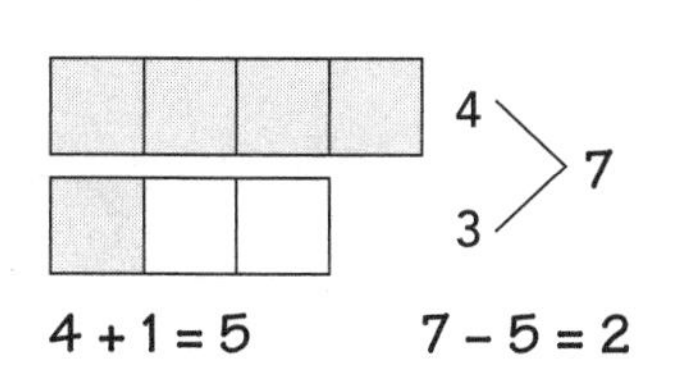

Most children find that Bar Modeling is very handy with multistep problems, particularly in keeping track of the various actions in the problem. The bar models serve as an easy reference and representation of these ideas.

Display Problem #66 on the interactive whiteboard.

King Kong had 8 mini-cupcakes. He ate 4 of them. Then he bought 5 more later. How many mini-cupcakes does King Kong have now?

King Kong has ___ mini-cupcakes now.

Read aloud the problem and have children underline the facts and circle the question on their papers.

Say: *Here we have another two-step problem. Two drawings may be helpful here as well. What is the first part of the problem?* (King Kong had 8 mini-cupcakes. He ate 4 of them.) *What kind of operation do we need to do?* (Subtraction) *Is it a take-away or comparison type of problem?* (Take away) *So how do we show this part of the problem?* (Draw a bar of 8 connected squares and cross out or shade in 4 squares.)

Have children draw the diagram on their paper, making sure they label it properly. Then have them write an equation to match the problem. Call on volunteers to share their diagrams and equations on the board (see Step 1 below).

Ask: *How many cupcakes are left now?* (4) *Now that we know that, what's the next part of the problem?* (King Kong bought 5 more cupcakes.) *What operation do we need to do here?* (Addition) *So how do we draw this?* (Draw two bars, one with 4 squares and the other with 5.)

Have children draw and label the diagram on their paper and write a matching equation. Then guide them through the process of solving the second part of the problem, which should look like Step 2 below.

STEP 1

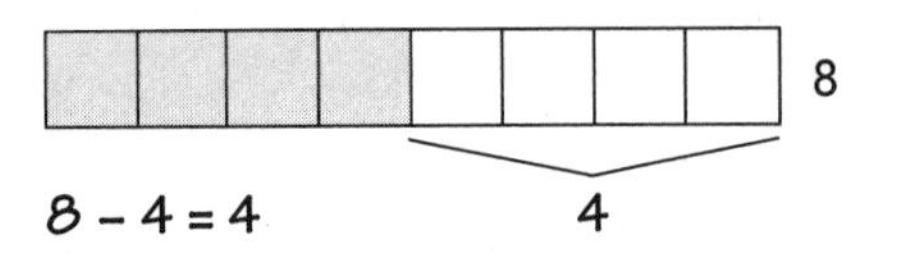

STEP 2

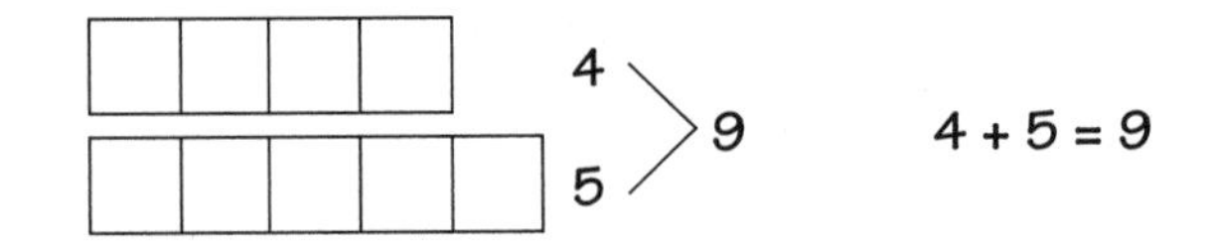

Another option is shown in the diagram below. We start with the addition (8 + 5) and then subtract the 4 to get 9. You may want to share this and guide children through this other process.

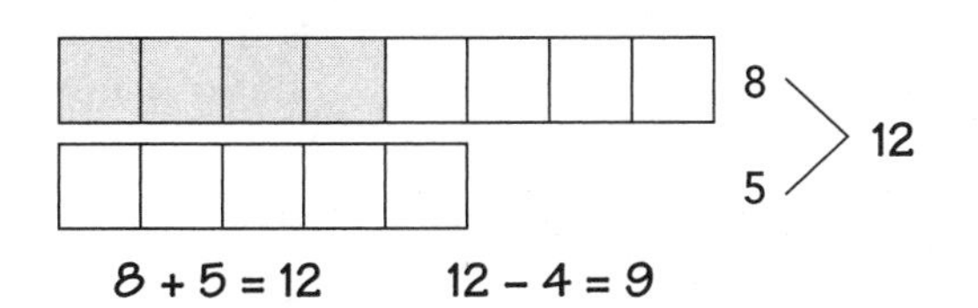

Have children work on Problems #67 and 68 (page 89) in pairs. Remind them to draw and label diagrams to represent the amounts in each problem. Have them write the equations too. Give children a few minutes to work. Then display each problem on the whiteboard. Call on volunteers to draw the diagrams and write the equations on the board. Then invite children to share their strategies for solving the problems, interacting on the whiteboard whenever possible.

Document Camera Sharing

A document camera or other device that can project student work can be very helpful during these problem discussions. Having children draw their diagrams on their papers and then share via the projection system saves time and honors student work in a very visible way.

LESSON 18

More Mixed Multistep Problems

Materials: student pages 90–91, pencils, projector, interactive whiteboard, markers

Preparation: Distribute copies of pages 90–91 and pencils to children. Go to www.scholastic.com/problemsolvedgr1 and click on Lesson 18. Set up your computer and projector to display the problems on the interactive whiteboard.

In this lesson children will encounter more multistep word problems that use both addition and subtraction.

Display Problem #69 on the interactive whiteboard.

Uncle Pete made 7 hot dogs. Uncle Bob made 8 hot dogs. The kids ate 10 of the hot dogs. How many hot dogs were left?

There were ___ hot dogs left.

Read aloud the problem to the class.

Say: *Here's another two-step problem. What are the facts?* (Uncle Pete made 7 hot dogs; Uncle Bob made 8. The kids ate 10 hot dogs.) *What is the question?* (How many hot dogs were left?) *What is our first step to solving this problem?* (Add Uncle Pete and Uncle Bob's hot dogs) *How should we show that?* (Draw two bars, one with 7 squares and one with 8.)

Have children draw the diagram on their papers, making sure they label it correctly and write a matching equation. Call on volunteers to share their diagrams and equations on the board, then discuss strategies for solving this part of the problem (see Step 1 below).

Ask: *How many hot dogs do we have now?* (15) *What is our next step?* (Subtract the 10 hot dogs the kids ate) *How do we show that?* (Draw a bar of 15 connected squares, then cross out or shade in 10 squares.)

Have children draw the diagram on their paper and solve the problem (see Step 2). Invite them to share their strategies and solutions on the board.

STEP 1

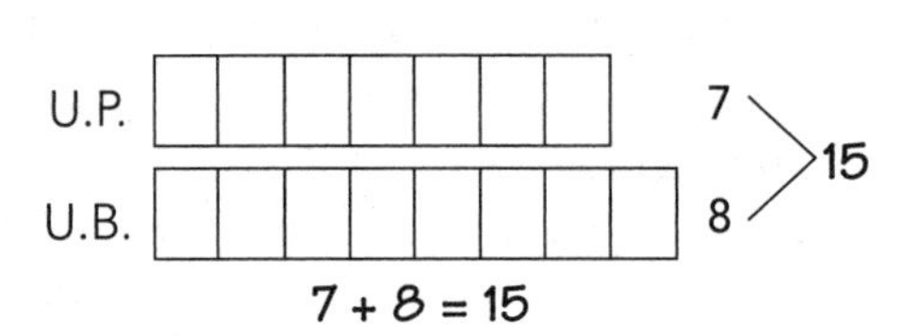

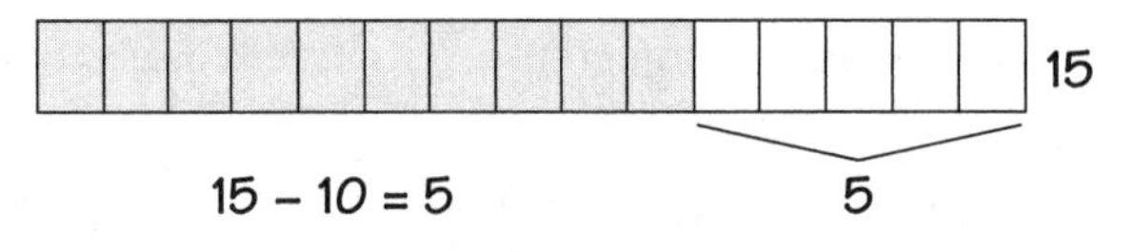

In this one-diagram strategy, we take the 10 that the kids ate and split it into two 5s. We can then take 5 from each quantity, so that the 5 remaining can easily be seen, as shown below.

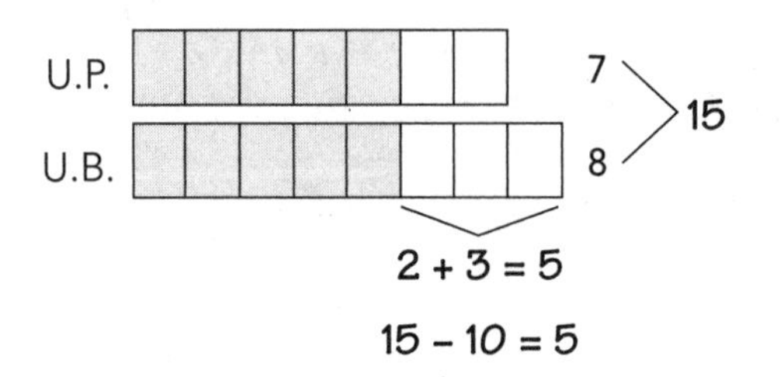

Display Problem #70 on the interactive whiteboard.

Read aloud the problem and have children underline the facts and circle the question on their papers.

Say: *This problem is all about addition. There are many ways we can show the addends. We can draw one bar of squares for each addend. If we do that, how many bars would we draw?* (4)

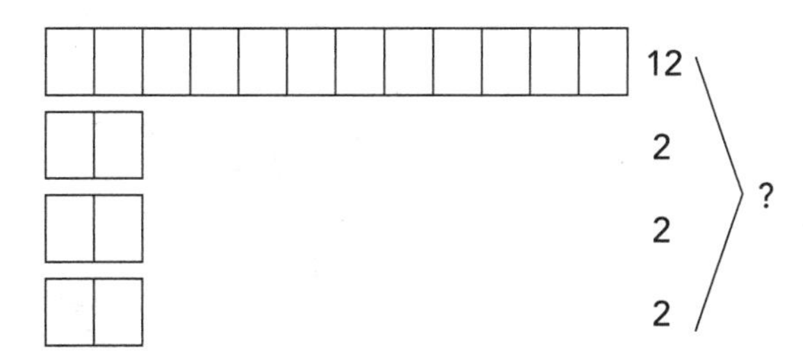

Say: *We can also combine some of the addends into one bar.* Demonstrate how we can take the three 2s and combine them into one bar to make 6. We can then add on 10 from the 12 bar plus the 2 extra, as shown below.

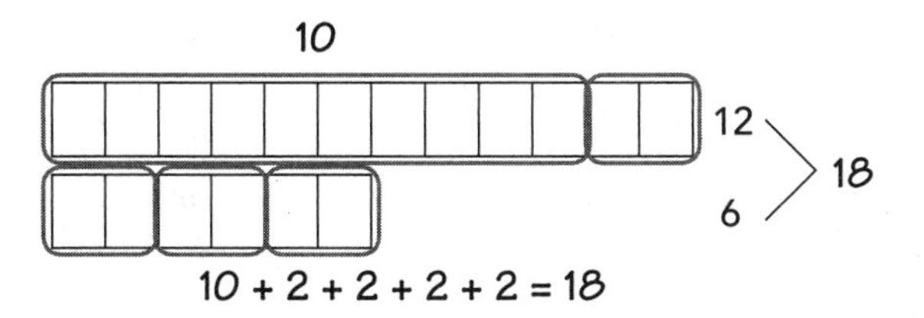

Have children work on Problems #71 and 72 (page 91) in pairs. Remind them to draw and label diagrams to represent the amounts in each problem. Have them write the equations too. Give children a few minutes to work. Then display each problem on the whiteboard. Call on volunteers to draw the diagrams and write the equations on the board. Then invite children to share their strategies for solving the problems, interacting on the whiteboard whenever possible.

LESSON 19

Mixed Multistep Problems With Larger Numbers

Materials: student pages 92–93, pencils, projector, interactive whiteboard, markers

Preparation: Distribute copies of pages 92–93 and pencils to children. Go to www.scholastic.com/problemsolvedgr1 and click on Lesson 19. Set up your computer and projector to display the problems on the interactive whiteboard.

The multistep problems in this lesson involve adding and subtracting larger numbers, as well as repeated addition. These are good problems to have children practice drawing continuous bars instead of connected squares. Most quantities are large enough so that drawing connected squares is time-consuming. You can transition to continuous bars by having children erase the lines between connected squares.

Display Problem #73 on the interactive whiteboard.

25 children came to school wearing hats. 10 hats were yellow. 5 were red. The rest were blue. How many hats were blue?

There were ___ blue hats.

Read aloud the problem to the class.

Ask: *What do we know in this problem?* (25 children wore hats. 10 hats were yellow. 5 were red. The rest were blue.) *What do we want to know?* (How many hats were blue?)

Say: *There are a few ways we can solve this problem. We can add the yellow and red hats first, then subtract the total from 25. Or we could start with the 25 hats and subtract the yellow hats from it, and then the red hats. The remainder will tell us how many hats were blue.*

Encourage children to try one of these strategies or come up with their own. Have them draw their diagrams on their papers, then call on volunteers to share their strategies and solutions on the board.

In the diagram on page 53, we start with a bar to represent the 25 hats. Then we shade in 10 squares to show the yellow hats. That leaves us with 15 hats. We then shade in another 5 to represent the red hats. The remaining unshaded section is 10—the number of blue hats.

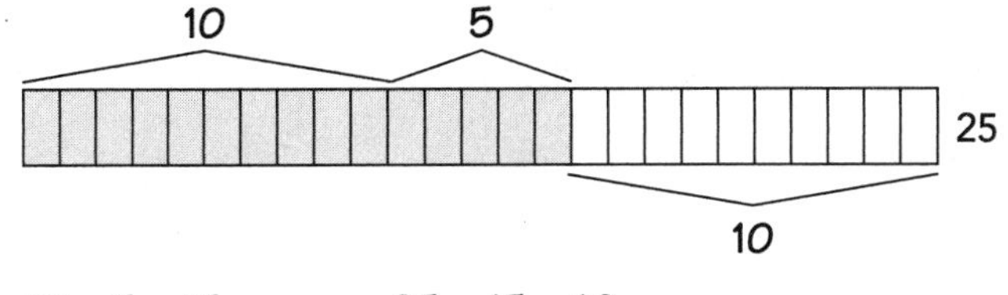

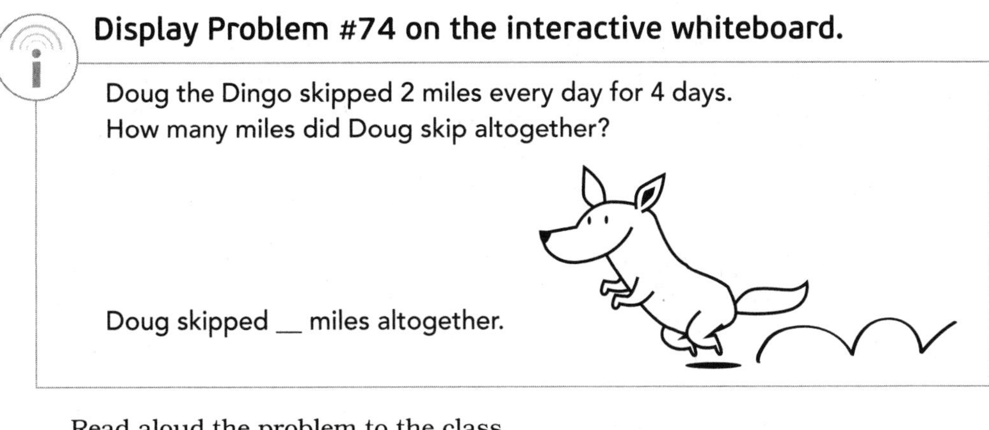

Read aloud the problem to the class.

Say: *In this problem the same amount is added again and again. Doug skipped 2 miles every day for 4 days. He skipped the same amount 4 times. We can draw four bars to show each day. But let me show you another way to do this.*

Draw one long bar on the board and divide it into four equal sections.

Explain: *We can draw one long bar that's divided into 4 sections. Each section is worth 2 miles. I'll draw the arrow and question mark here to show that we are trying to find out how many the sections are worth altogether.*

Have children work in pairs to solve the problem. Then invite volunteers to share their solutions on the board. An easy way to find the answer is to skip count by 2s to reach 8.

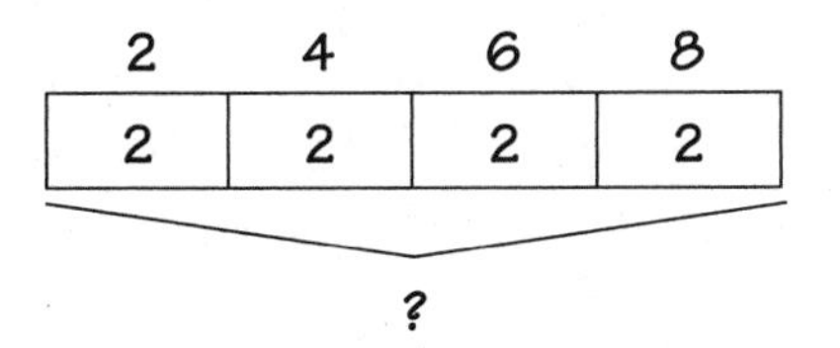

Have children work on Problems #75 and 76 (page 93) in pairs. Remind them to draw and label diagrams to represent the amounts in each problem. Have them write the equations too. Give children a few minutes to work. Then display each problem on the whiteboard. Call on volunteers to draw the diagrams and write the equations on the board. Then invite children to share their strategies for solving the problems, interacting on the whiteboard whenever possible.

100s Grid

Many problems involving comparisons of numbers can also be represented on a 100s grid or chart. You may want to show children how this tool can also be used to represent quantities and strategies in problem solving. This is a good connection to help children see. It lets them know that the same ideas can be represented using a variety of math tools and strategies.

LESSON 20

Challenging Multistep Problems

Materials: student pages 94–95, pencils, projector, interactive whiteboard, markers

Preparation: Distribute copies of pages 94–95 and pencils to children. Go to www.scholastic.com/problemsolvedgr1 and click on Lesson 20. Set up your computer and projector to display the problems on the interactive whiteboard.

Children will encounter more challenging problems with larger numbers and multiple steps. They will find Bar Modeling helpful as a tool or strategy for solving the problems and as a way to show thinking.

Display Problem #77 on the interactive whiteboard.

Tangy spent $15 on a new toy mushroom. He only has $19 left.
How much money did Tangy start with?

Tangy started with ___ dollars.

Read aloud the problem. This can be a challenging problem for children to understand because the first sentence discusses money being spent yet there is no subtraction to do. Instead we are trying to find what Tangy started with.

Ask: *What do we know in this problem?* (Tangy spent $15 on a new toy mushroom. He has only $19 left.) *What do we want to find out?* (How much money did Tangy start with?) *What operation do we need to use to find the answer?* (Addition)

Explain: *Usually when we see a problem where something was spent, we think of subtraction. But this problem tells us how much Tangy spent and how much he has left. We want to know how much he started with. To do this, we need to add what he spent and what was left. Let's draw some bars to show what we know.*

Guide children to draw two bars on their paper to show the two amounts. Remind them to label the bars, then to add the arrow and question mark so we know what we're looking for. Have them write a matching equation as well. Then call on volunteers to share their diagrams and equations on the board.

Encourage children to work in pairs to come up with strategies for solving the problem. Then invite them to share their strategies and solutions on the board.

Here's one strategy you could share with the class:

Explain: *Here we have two amounts, 15 and 19, that we need to add. These are pretty big numbers. We can look for "friendly numbers" to work with by sectioning the bars. For example, we can start by sectioning the bars so we can see 10s and 5s. We know that 15 is 10 + 5, and that 19 is 10 + 9. We can also section 19 to*

10 + 5 + 4. Now we can combine the 10s: 10 + 10 = 20. We can also combine the 5s: 5 + 5 = 10. That makes 30. Add the extra 4, and we have 34. Tangy started with $34. Share the diagram below on the board.

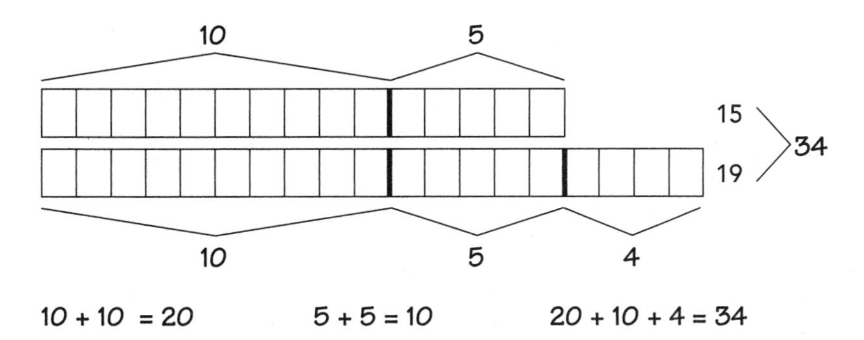

Display Problem #78 on the interactive whiteboard.

Lois has three sticks that total 25 feet long. The first stick is 7 feet long. The second stick is 9 feet long. How long is the third stick?

The third stick is ___ feet long.

Read aloud the problem to the class.

Say: *Here's another problem we can solve different ways. Let's start with what we know. We know that Lois has three sticks that are 25 feet long altogether. The first stick is 7 feet long. The second stick is 9 feet long. What do we need to find out?* (How long is the third stick?) *How can we show this?*

Encourage children to share their strategies for solving this problem using Bar Modeling. There are many approaches to this problem. One way would be to add the lengths of the first and second sticks, then subtract the total from 25. Another way is to represent the 25 with one bar and then cut back 7 and again cut back 9, subtracting twice. If we then section off 4 more, that makes 20. The last section must be 5 and 5 + 4 = 9. (See below.)

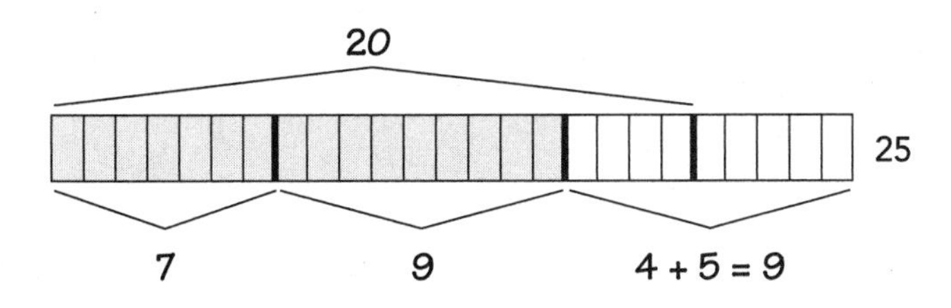

Have children work on Problems #79 and 80 (page 95) in pairs. Remind them to draw and label diagrams to represent the amounts in each problem. Have them write the equations too. Give children a few minutes to work. Then display each problem on the whiteboard. Call on volunteers to draw the diagrams and write the equations on the board. Then invite children to share their strategies for solving the problems, interacting on the whiteboard whenever possible.

Name ______________________________

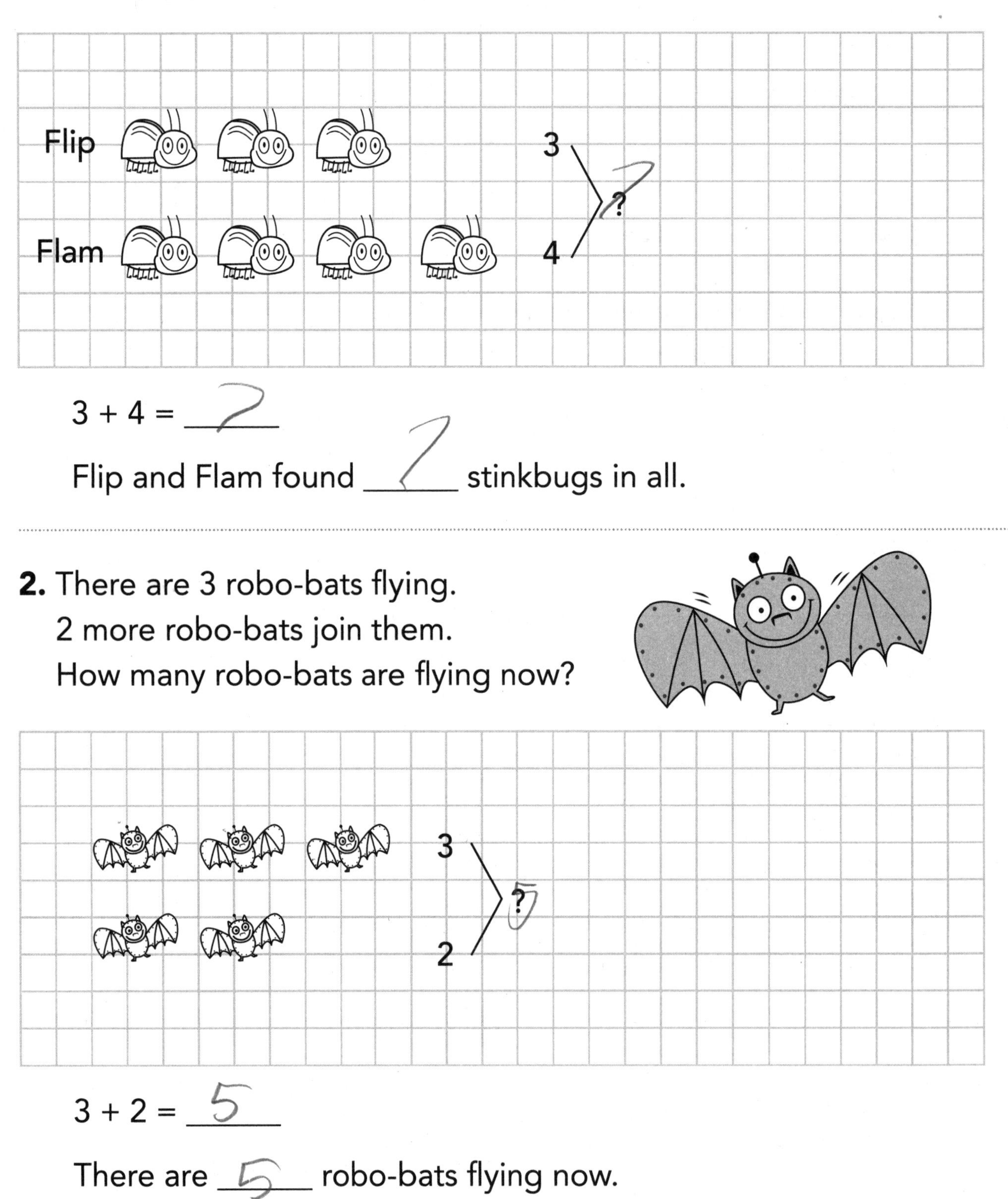

1. Flip found 3 stinkbugs. Flam found 4 stinkbugs.
How many stinkbugs did they find in all?

3 + 4 = 7

Flip and Flam found 7 stinkbugs in all.

2. There are 3 robo-bats flying.
2 more robo-bats join them.
How many robo-bats are flying now?

3 + 2 = 5

There are 5 robo-bats flying now.

Name ______________________________

3. Jill ate 4 bongo berries. Then she ate 4 more bongo berries. How many bongo berries did she eat altogether?

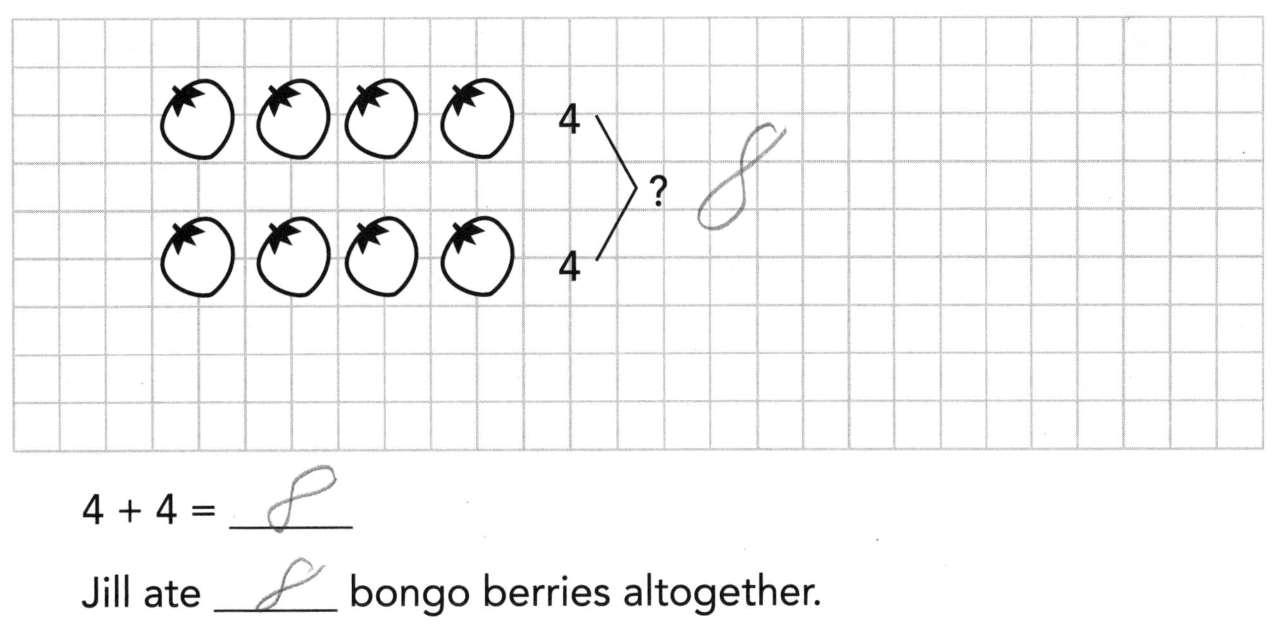

4 + 4 = __8__

Jill ate __8__ bongo berries altogether.

---

4. Tim painted 3 spaceships green. He painted 1 spaceship red. How many spaceships did Tim paint in all?

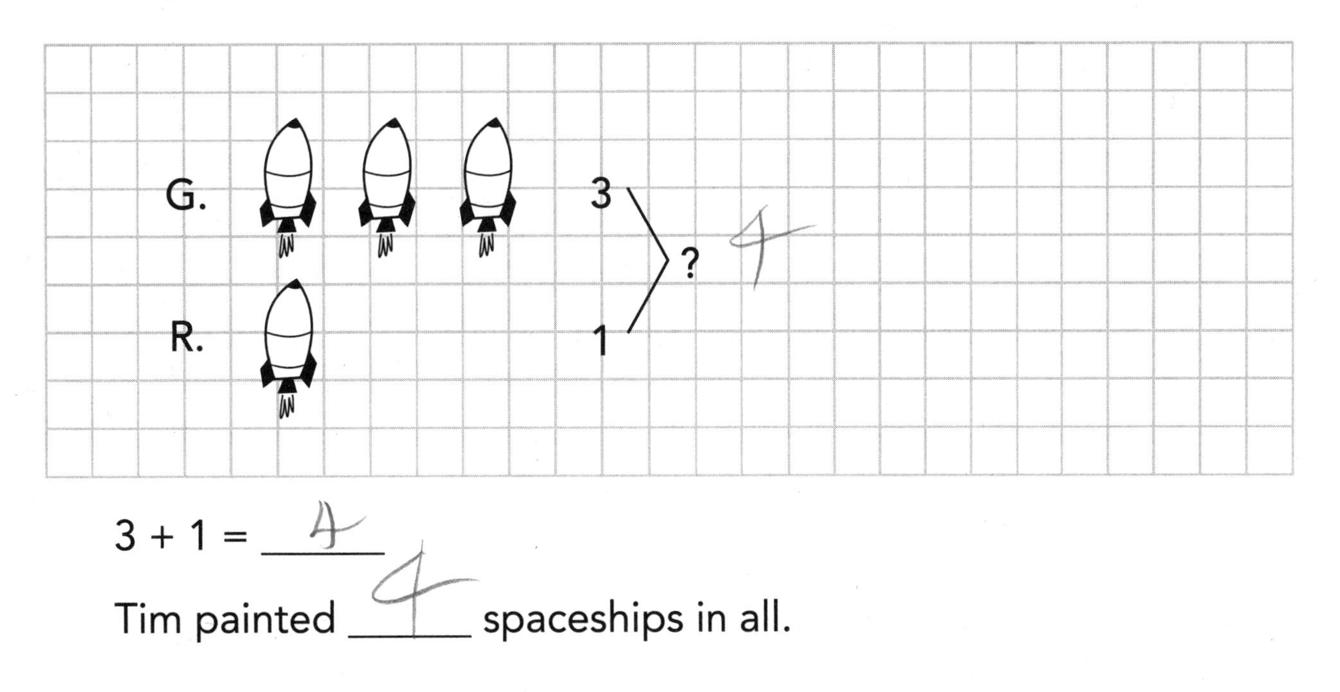

3 + 1 = __4__

Tim painted __4__ spaceships in all.

Name ____

5. Maxie ate 2 super cookies. His sister Flaxie ate 5 super cookies. How many super cookies did they eat altogether?

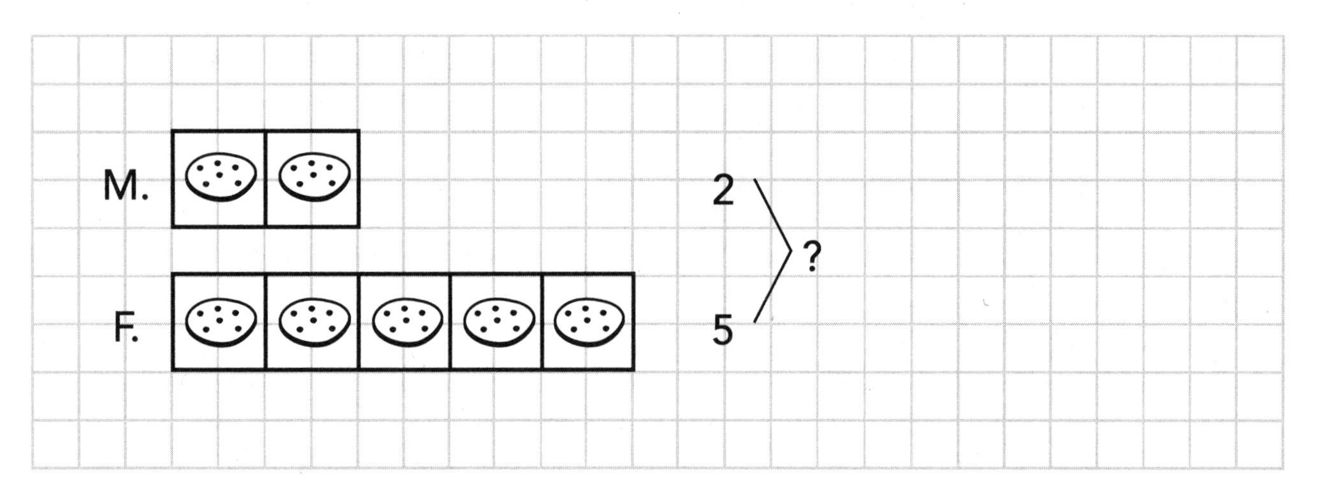

2 + 5 = ______

Maxie and Flaxie ate ______ super cookies altogether.

---

6. There were 6 birds sitting in a nest. Then 1 more bird joined them. How many birds are in the nest now?

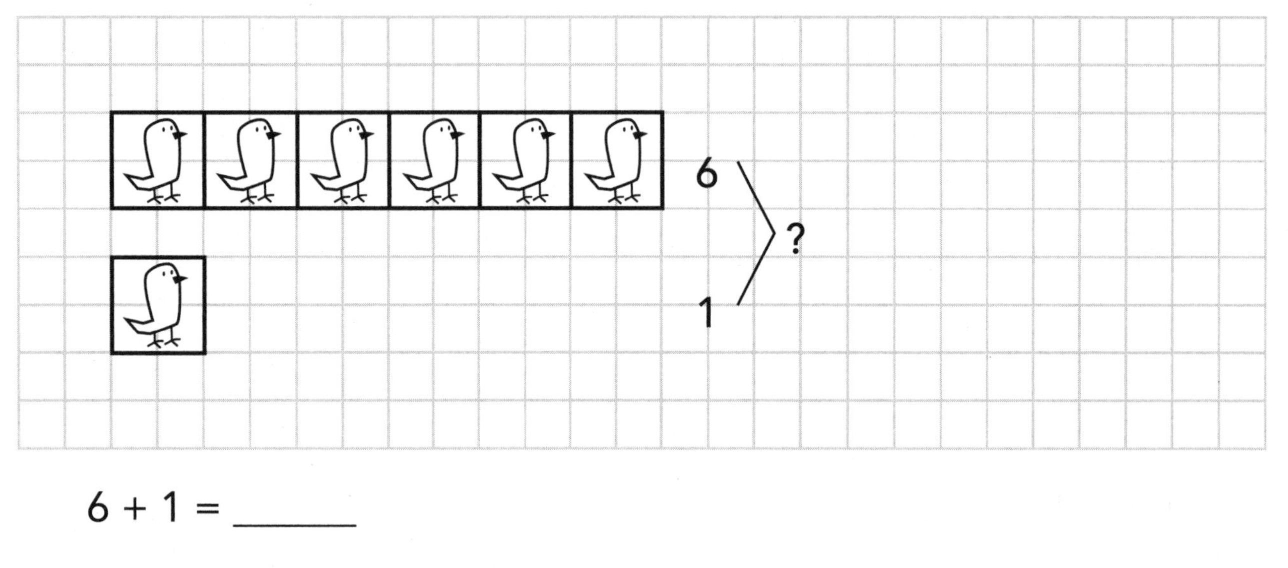

6 + 1 = ______

There are ______ birds in the nest now.

Name ______________________________

7. Alan the ape had 3 coconuts. He picked 3 more. How many coconuts does Alan have altogether?

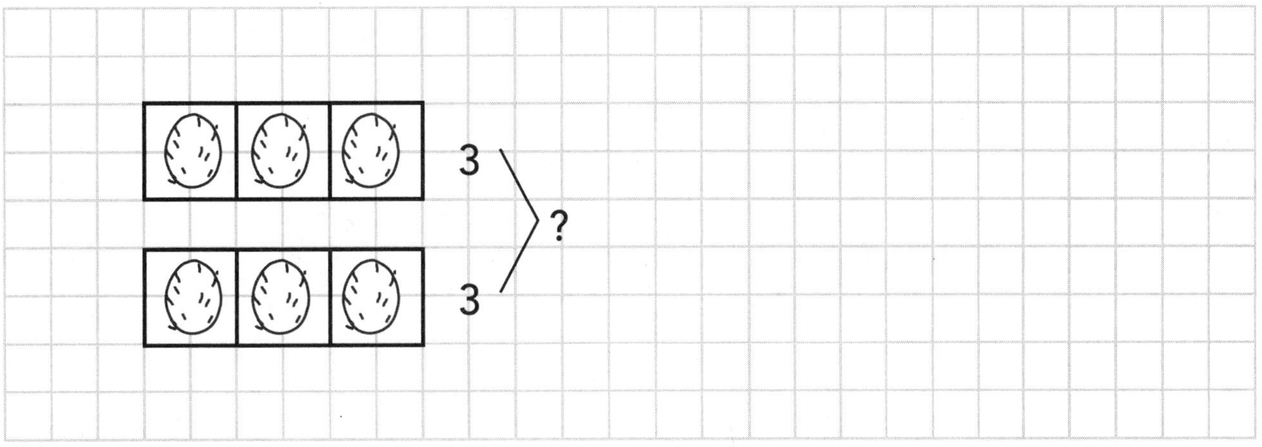

3 + 3 = ______

Alan has ______ coconuts altogether.

8. There are 6 dogs in the tub. There are 2 dogs in the sink. How many dogs are there in all?

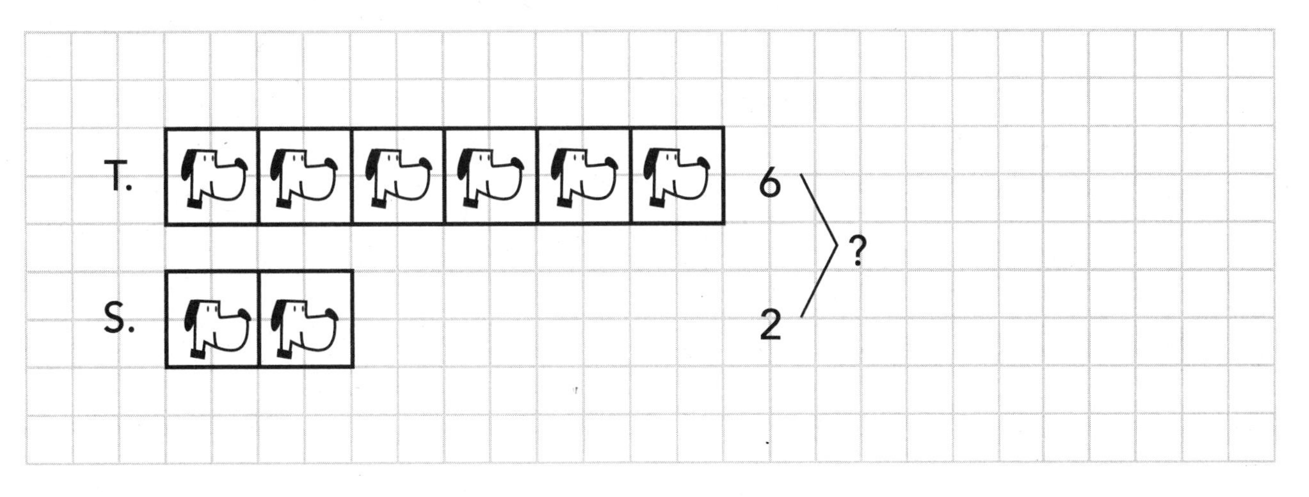

6 + 2 = ______

There are ______ dogs in all.

Name ____

9. Stan caught 5 Giganto-Fish. Fran also caught 5 Giganto-Fish. How many Giganto-Fish did they catch in all?

S. 5

?

F. 5

5 + 5 = ______

Stan and Fran caught ______ Giganto-Fish in all.

10. Billy Worm counted 4 birds. Willy Worm counted 5 birds. How many birds did they count altogether?

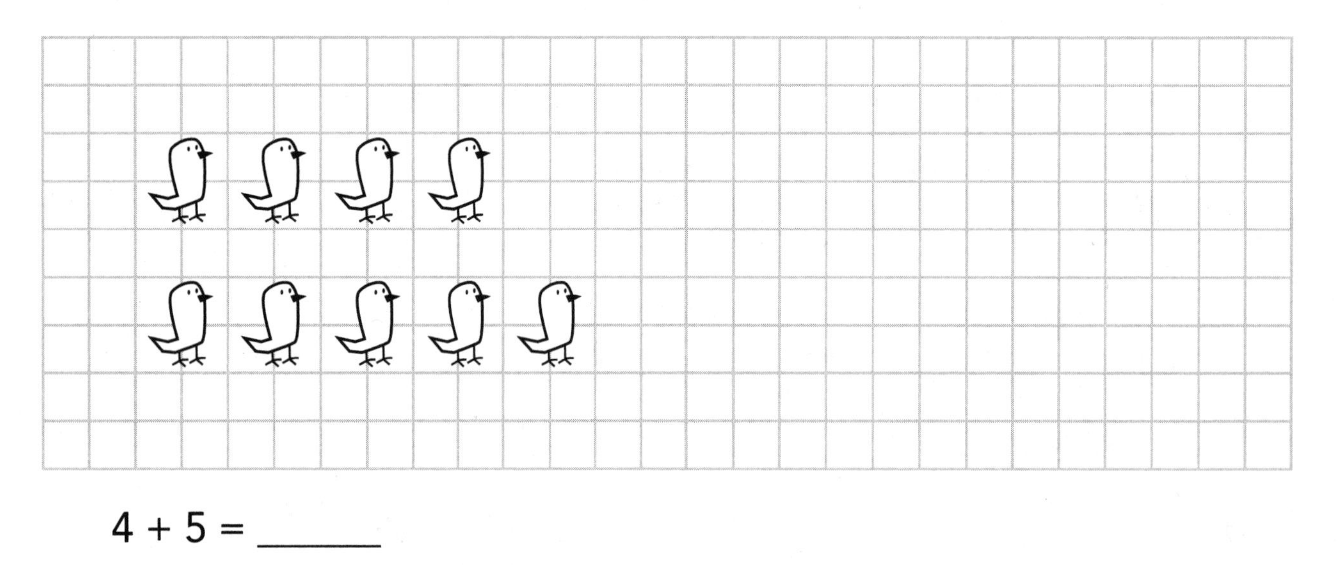

4 + 5 = ______

They counted ______ birds altogether.

Name ____________________

11. The teacher found 6 raccoons in Tom's desk.
She found 4 raccoons in Jan's desk.
How many raccoons did the teacher find?

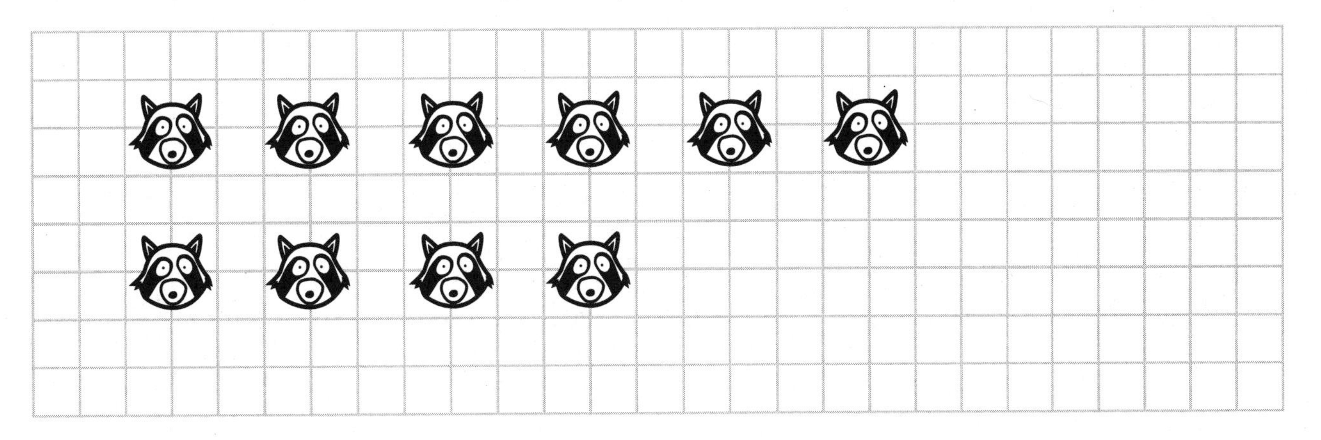

6 + 4 = ______

The teacher found ______ raccoons.

12. Flip wanted to get hats for all his friends.
He had 4 friends that were girls and
6 friends that were boys.
How many hats does Flip need?

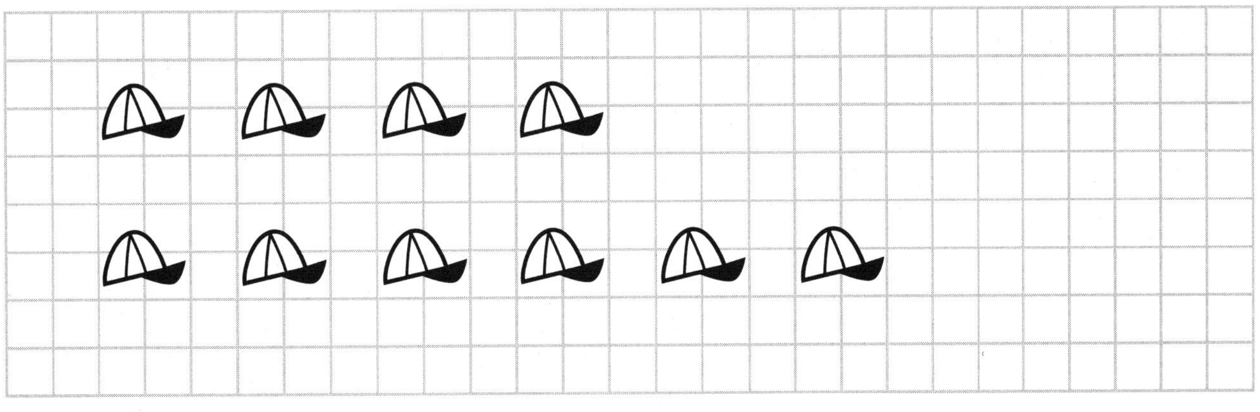

4 + 6 = ______

Flip needs ______ hats.

Name ____________________

13. Pete has 5 glow socks.
Paul has 4 glow socks.
How many glow socks
do the boys have in all?

5 + 4 = ______

The boys have ______ glow socks in all.

14. Pinky the poodle has 7 blue dog collars
and 3 green dog collars.
How many dog collars does Pinky have in all?

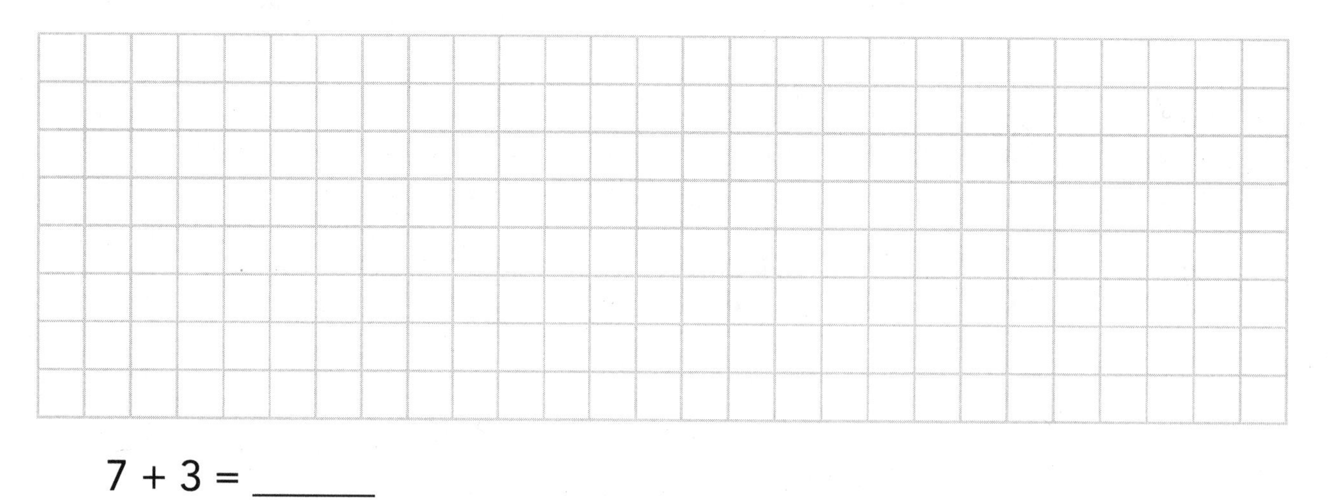

7 + 3 = ______

Pinky has ______ dog collars in all.

Name ______________________________

15. Breezy flew 9 kites. Sneezy flew 1 kite.
How many kites did they fly altogether?

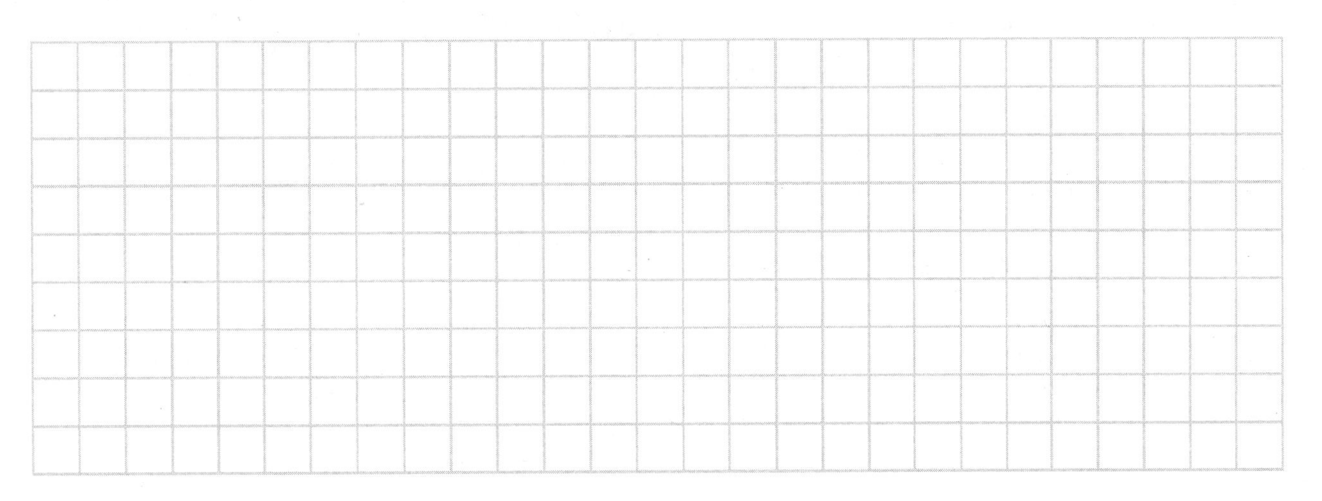

$9 + 1 =$ ______

They flew ______ kites altogether.

16. Bob the bear ate 8 bowls of berry cereal for breakfast. He ate 2 bowls for lunch.
How many bowls of berry cereal did Bob eat?

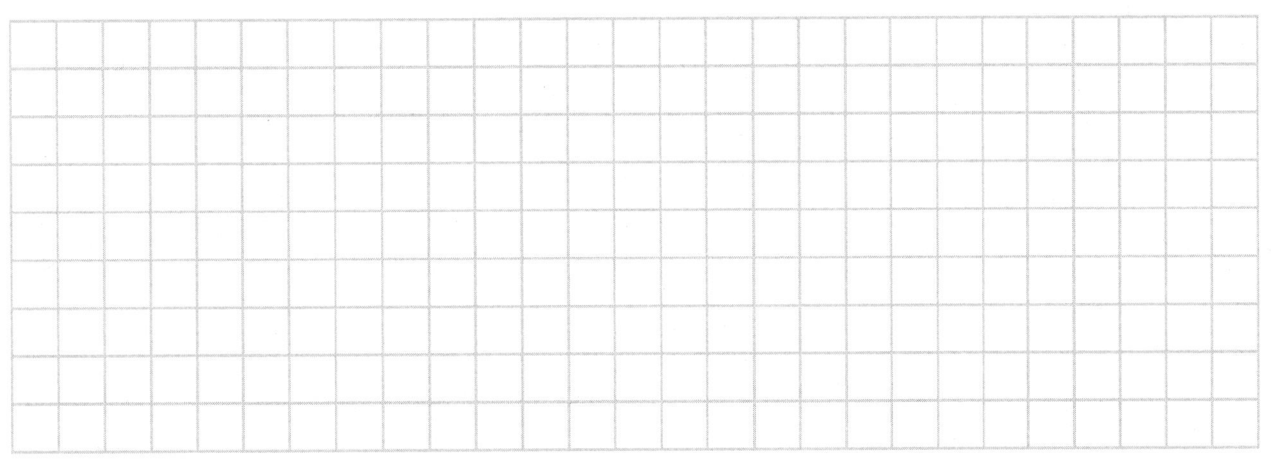

$8 + 2 =$ ______

Bob ate ______ bowls of berry cereal.

Name ____

17. Zip had 4 moon rocks. He threw 2 moon rocks at the Goo Goo Monster. How many moon rocks did Zip have left?

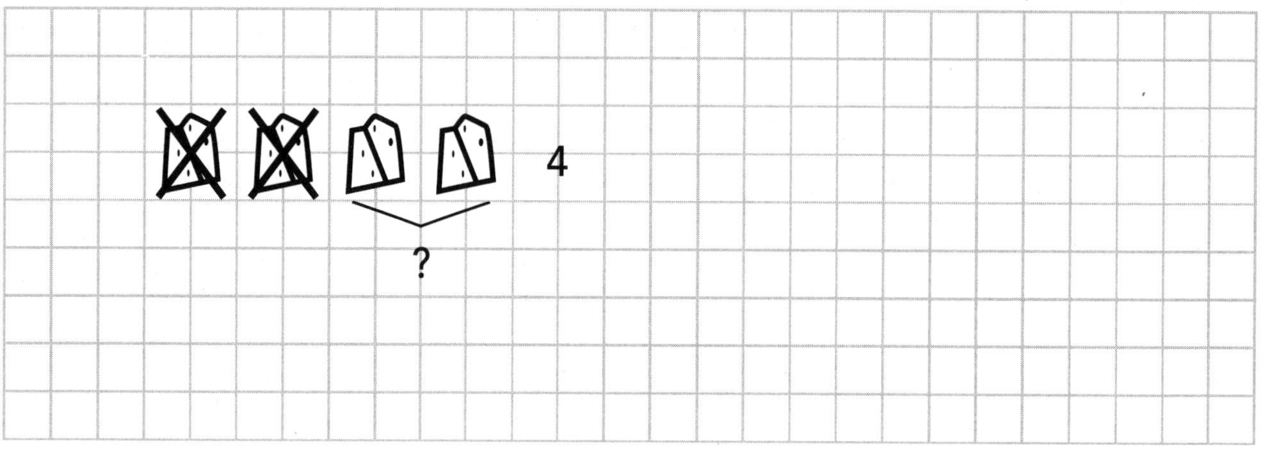

$4 - 2 =$ ______

Zip had ______ moon rocks left.

18. There were 6 space rockets. 1 space rocket took off. How many space rockets are left?

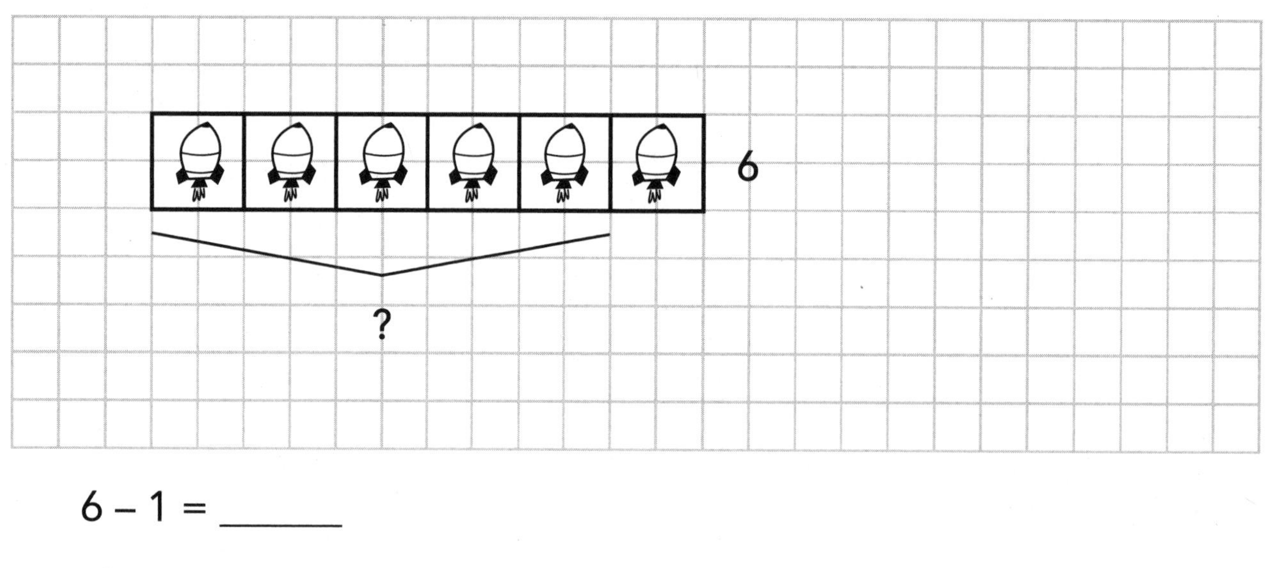

$6 - 1 =$ ______

There are ______ space rockets left.

Name ____

19. Tiny had 10 noodles in his pocket. 4 noodles fell out.
How many noodles does he have now?

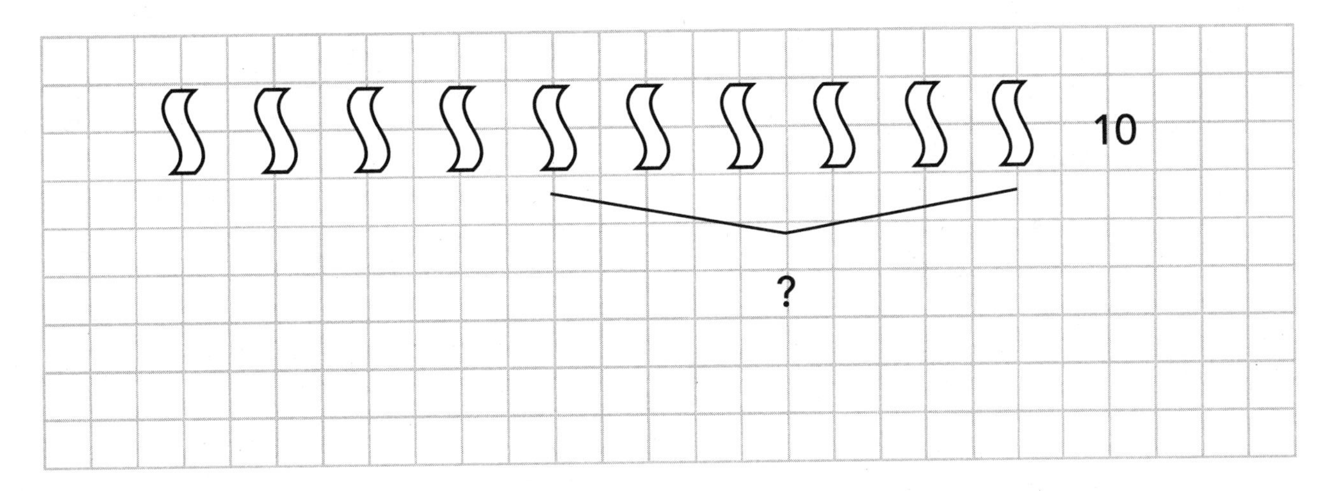

10 – 4 = ______

Tiny has ______ noodles now.

20. Fuzzy had 9 buggy bars.
He ate 3 buggy bars.
How many buggy bars are left?

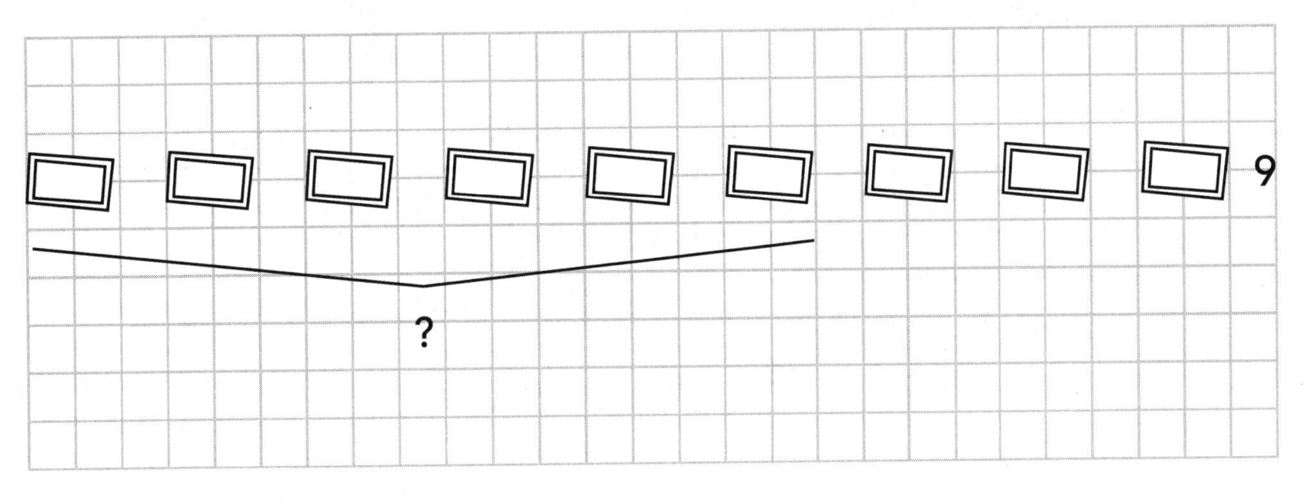

9 – 3 = ______

There are ______ buggy bars left.

Name ______________________________

21. Pat has 8 dino-hats. She has 6 cat-hats.
How many more dino-hats are there than cat-hats?

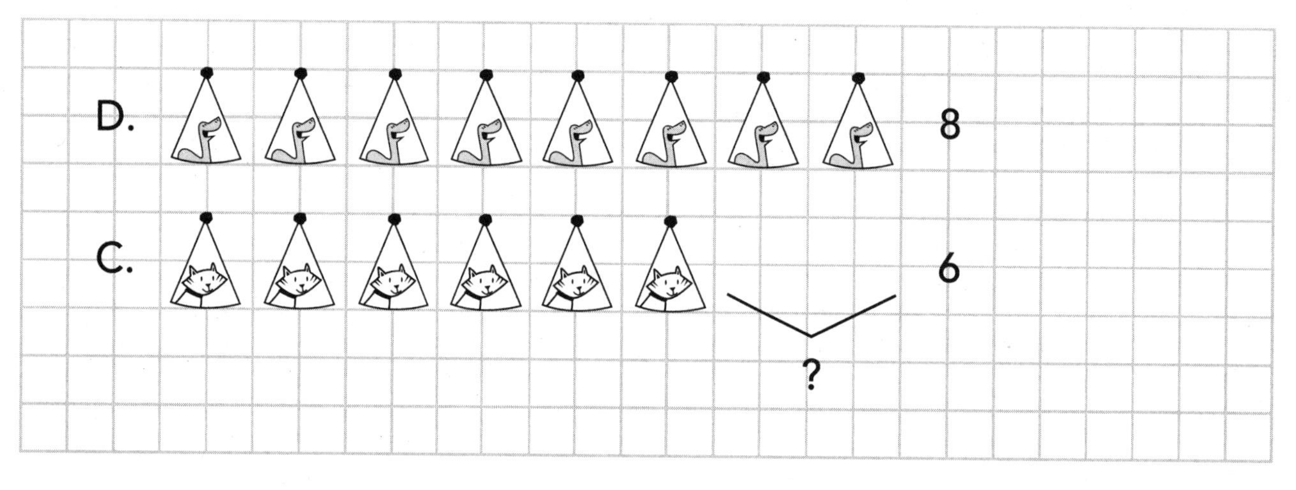

8 – 6 = ______

There are ______ more dino-hats than cat-hats.

22. Bizz has 4 coco-cookies.
Buzz has 10 coco-cookies.
What is the difference between the numbers of their coco-cookies?

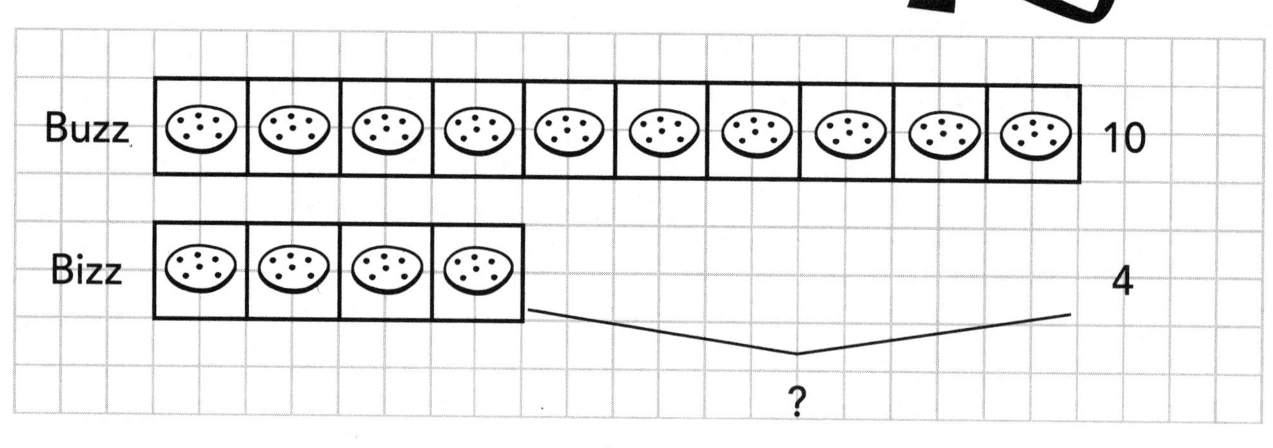

10 – 4 = ______

The difference between the numbers of coco-cookies is ______.

Name ____

23. Sid has 10 bonbons. Kerry has 3.
How many more bonbons does Sid have than Kerry?

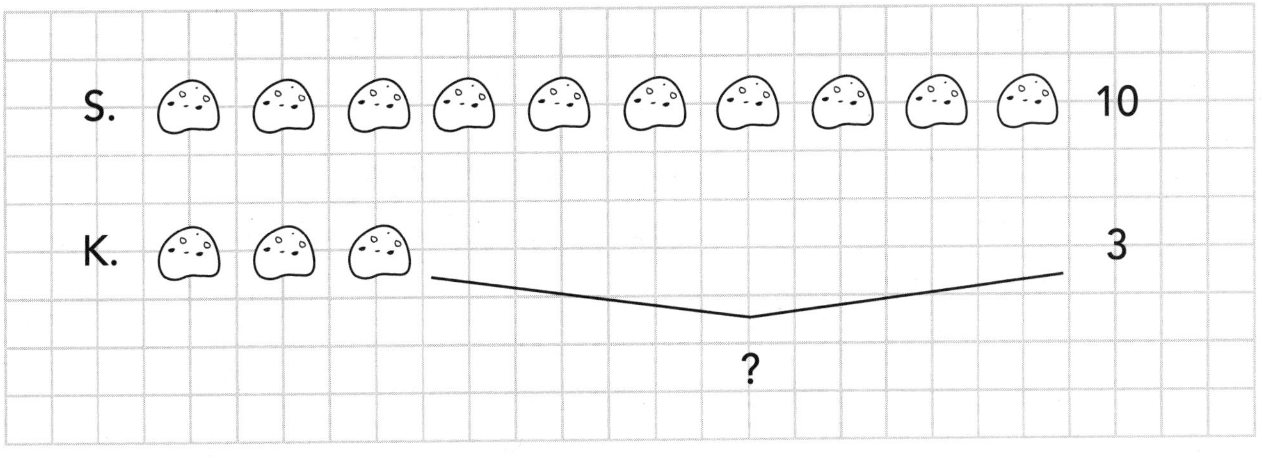

10 – 3 = ______

Sid has ______ more bonbons than Kerry.

24. Slick has a bag of gum. 6 pieces of gum are blue.
4 pieces of gum are white. How many
more blue pieces are there than white?

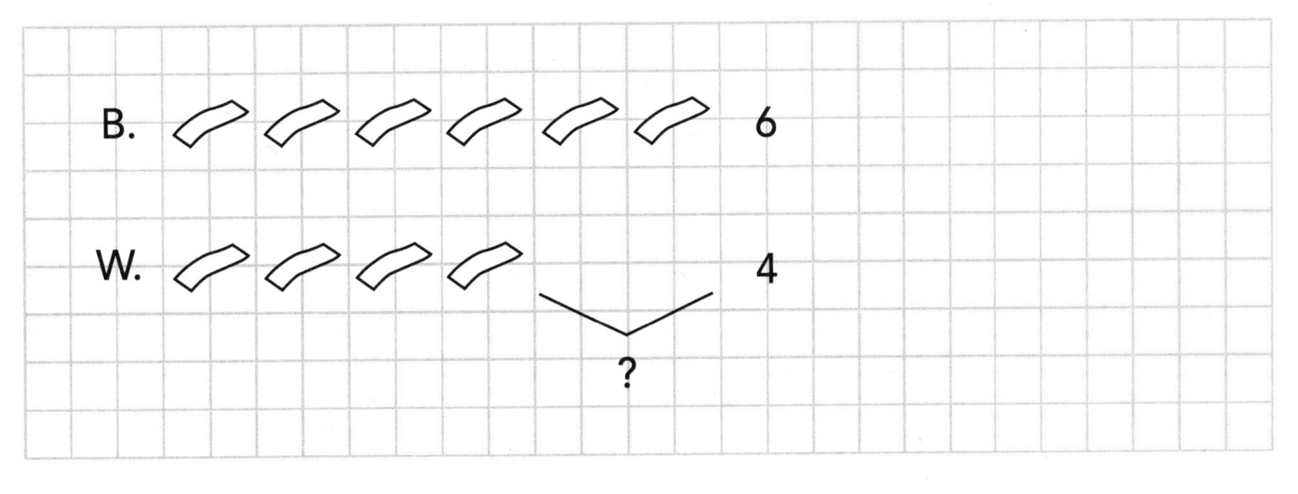

6 – 4 = ______

There are ______ more blue pieces than white.

Name ____________________

25. Bobo made 9 doggie cupcakes.
He sold 4 of them.
How many does he have left?

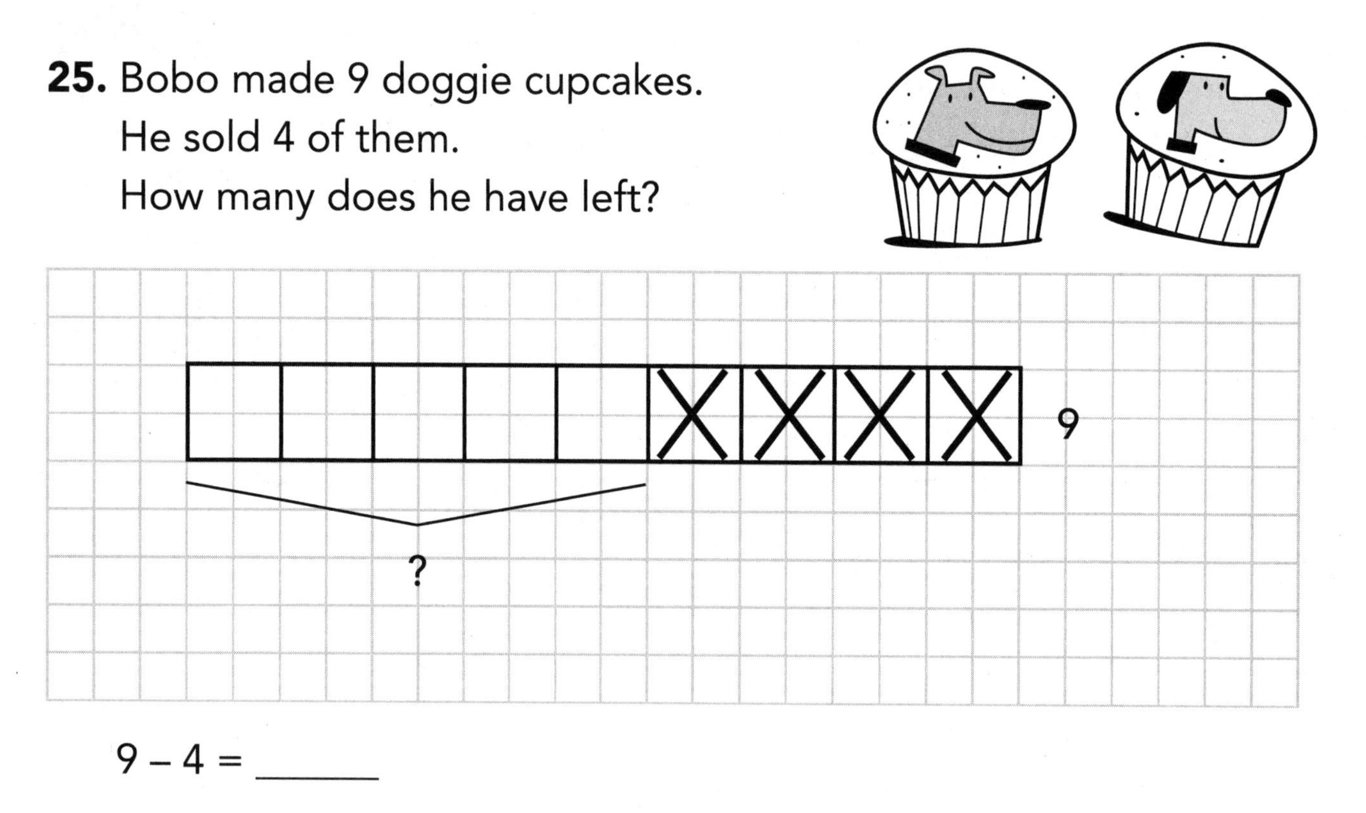

9 – 4 = ______

Bobo has ______ doggie cupcakes left.

26. Ipso had 10 peanuts. Bipso had 2 peanuts.
What is the difference between their numbers of peanuts?

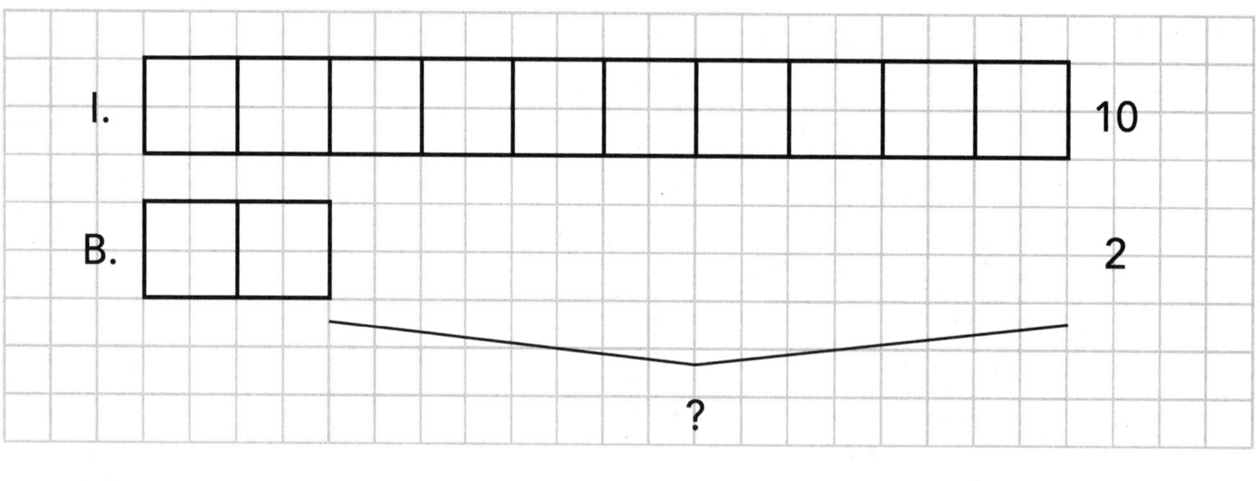

10 – 2 = ______

The difference between the numbers of peanuts is ______.

Name ______________________________

27. Stan has 10 toe rings. He lost 6 of them.
How many toe rings does Stan have left?

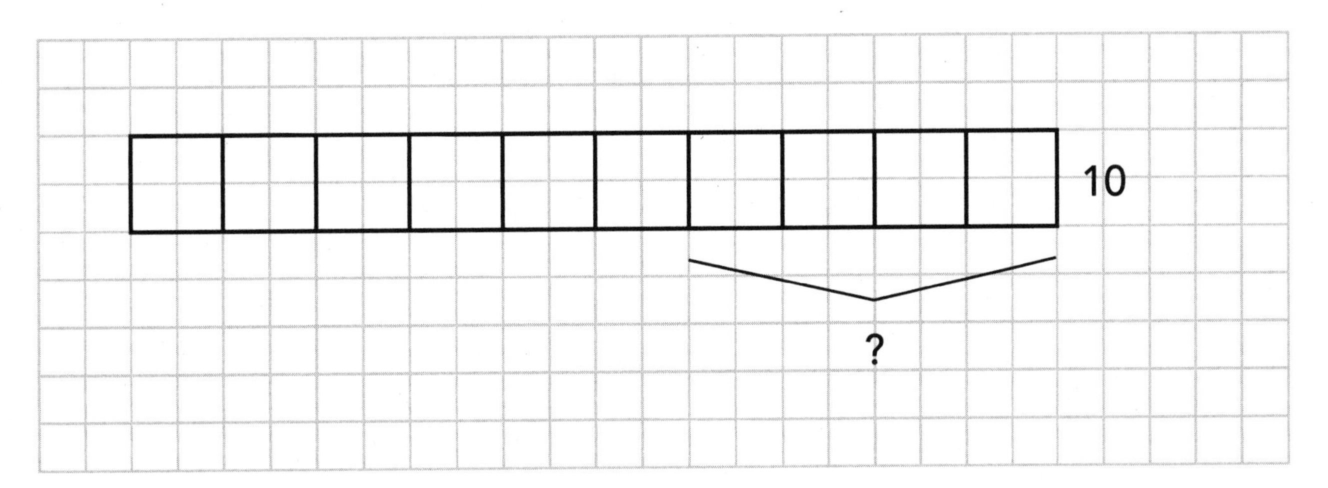

$10 - 6 =$ ______

Stan has ______ toe rings left.

28. Penny has 12 blue horns. She has 5 green horns. How many more blue horns are there than green?

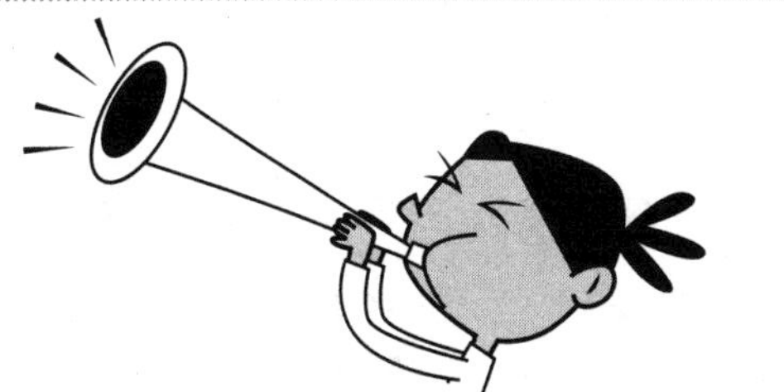

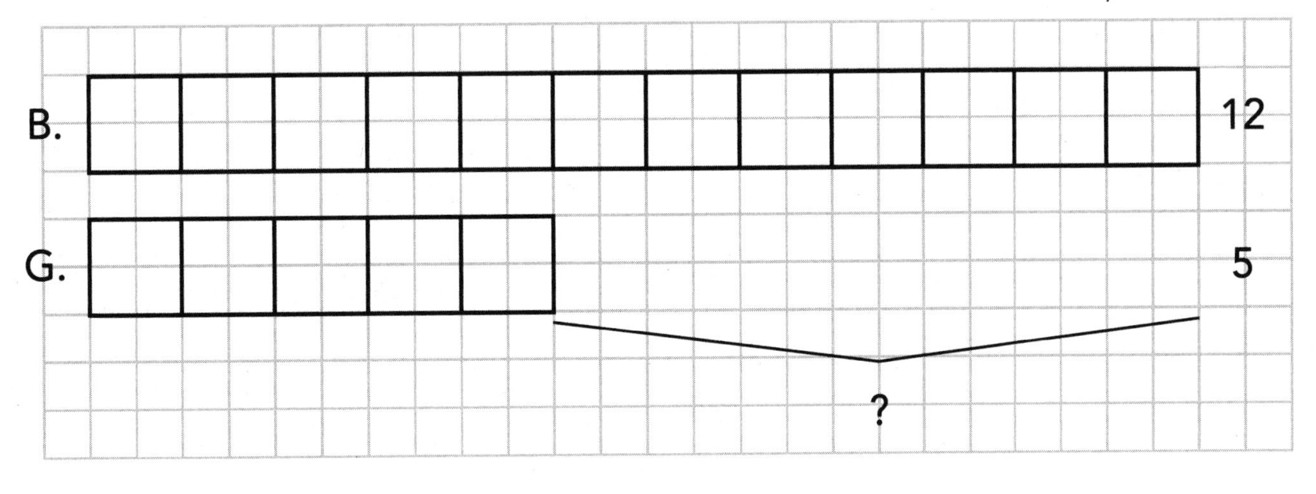

$12 - 5 =$ ______

There are ______ more blue horns than green.

Name ______________________________

29. Pippy had 9 pipsicle sticks. She gave 3 to her sister. How many pipsicle sticks does Pippy have left?

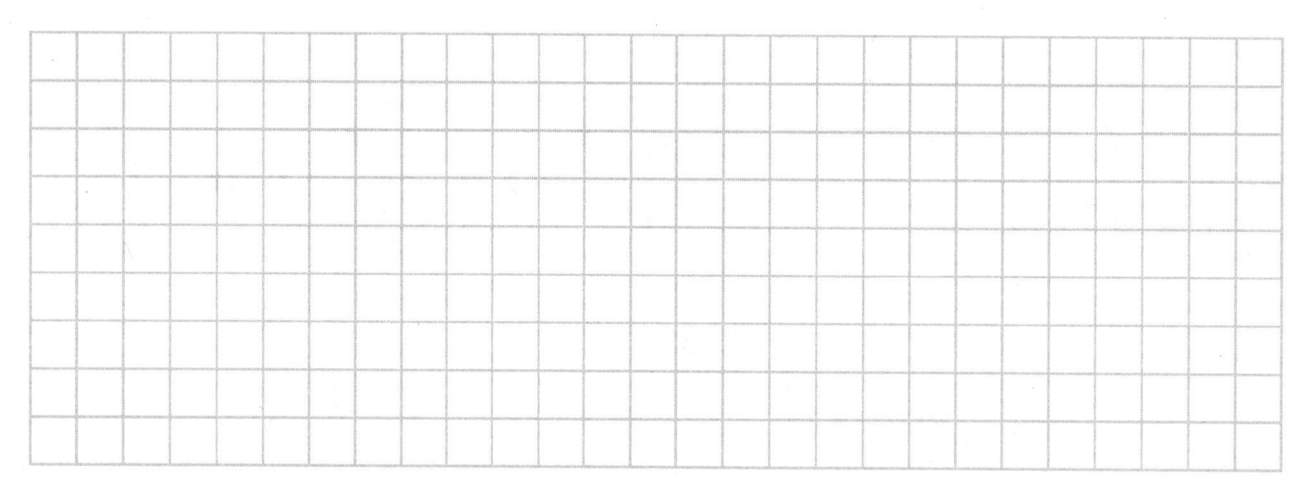

9 – 3 = ______

Pippy has ______ pipsicle sticks left.

---

30. Pina has 7 air donuts. Tina has 3 air donuts. How many more air donuts does Pina have than Tina?

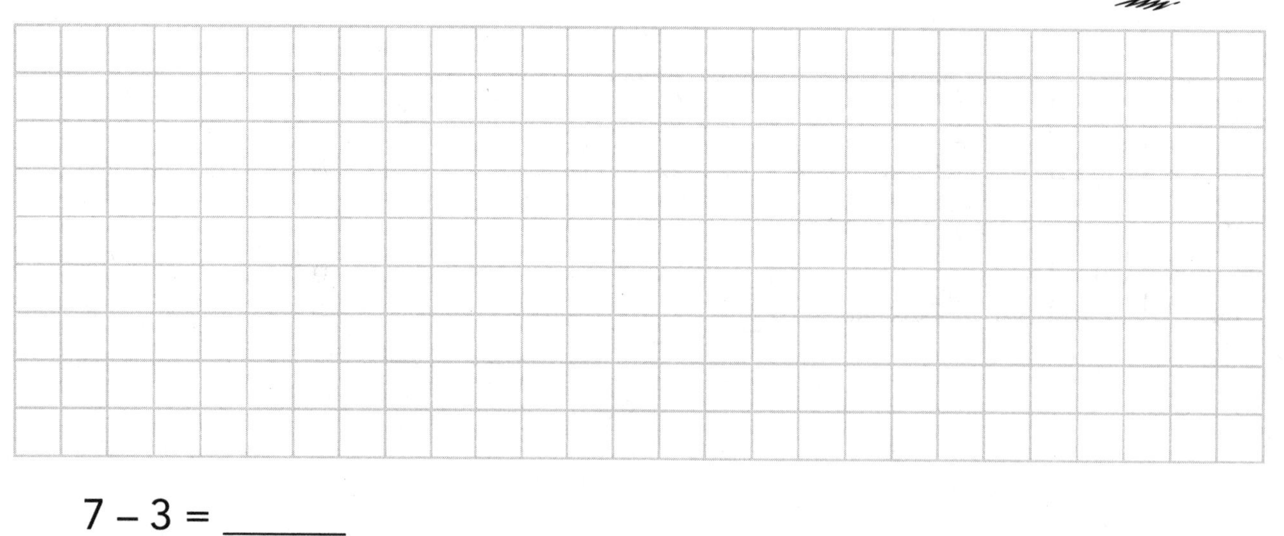

7 – 3 = ______

Pina has ______ more air donuts than Tina.

Name ____

31. Wooly Bully had 10 balls of yarn.
He colored 5 balls purple.
He colored the others yellow.
How many yellow balls of yarn were there?

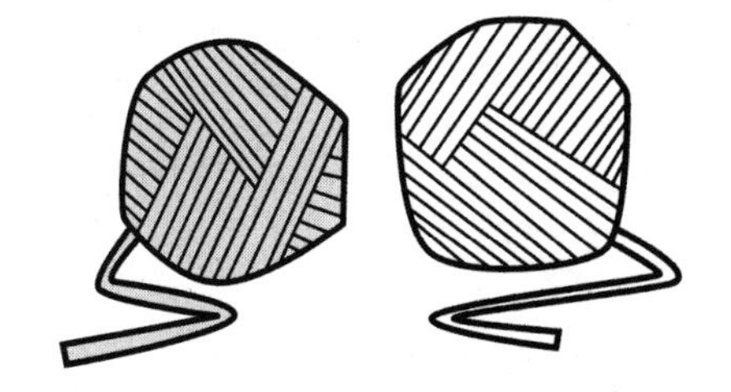

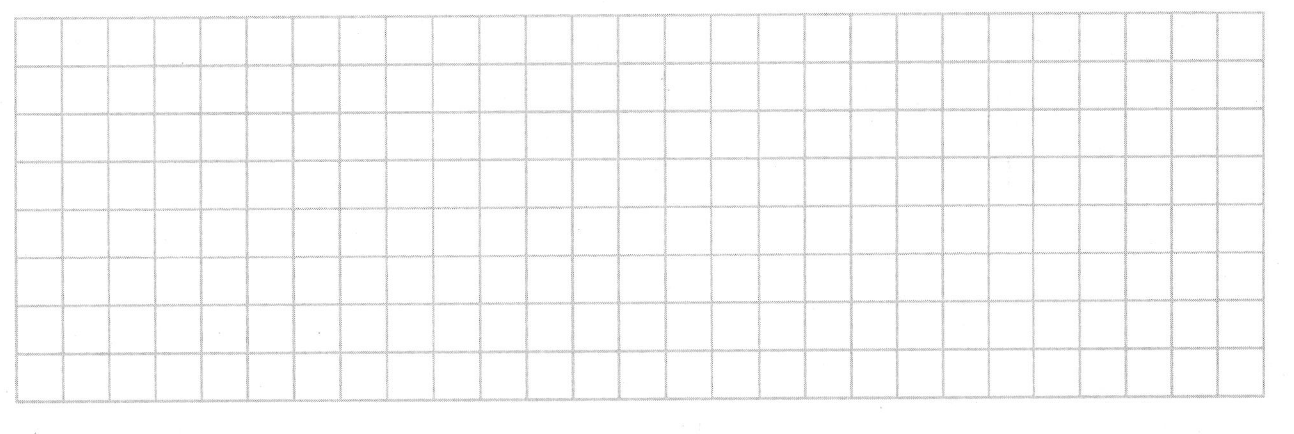

$10 - 5 =$ ______

There were ______ yellow balls of yarn.

32. There were 11 robot bats in their cosmic cave.
6 flew out. How many robot bats are still in the cave?

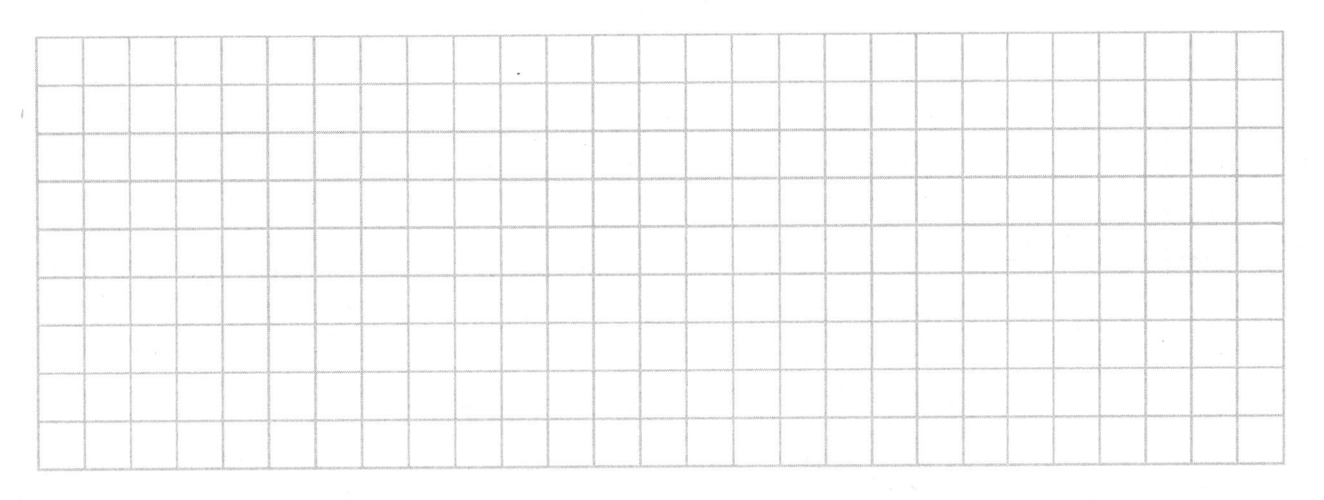

$11 - 6 =$ ______

There are ______ robot bats still in the cave.

Name ____

33. Smedley had 6 Zorbo balls. He bought 7 more at the store. How many Zorbo balls does he have now?

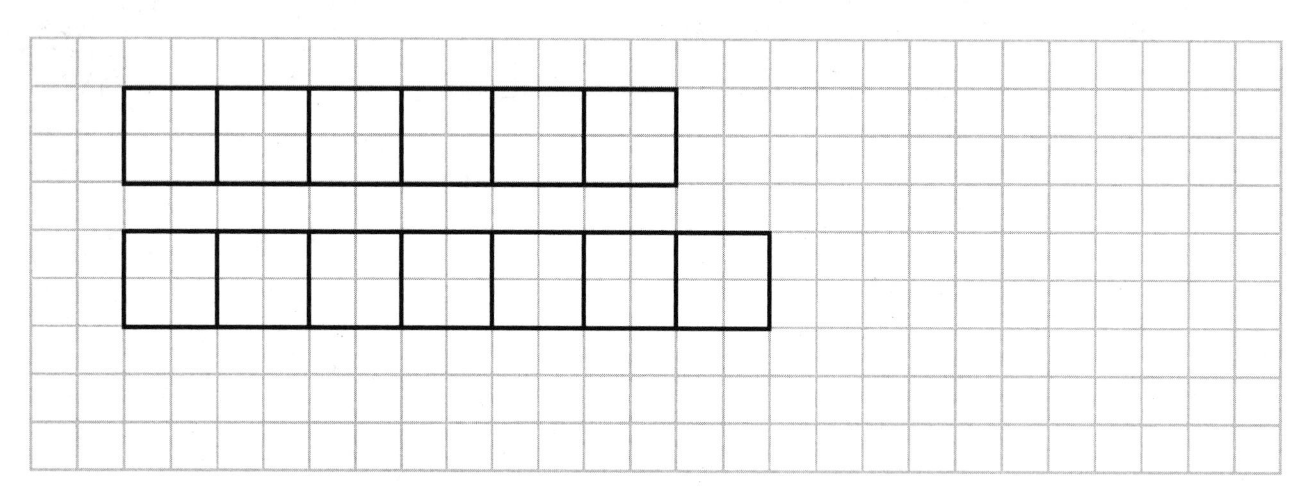

6 + 7 = ______

Smedley has ______ Zorbo balls now.

34. Ping made 13 purple pancakes. Pong made 5 purple pancakes. How many purple pancakes did they make altogether?

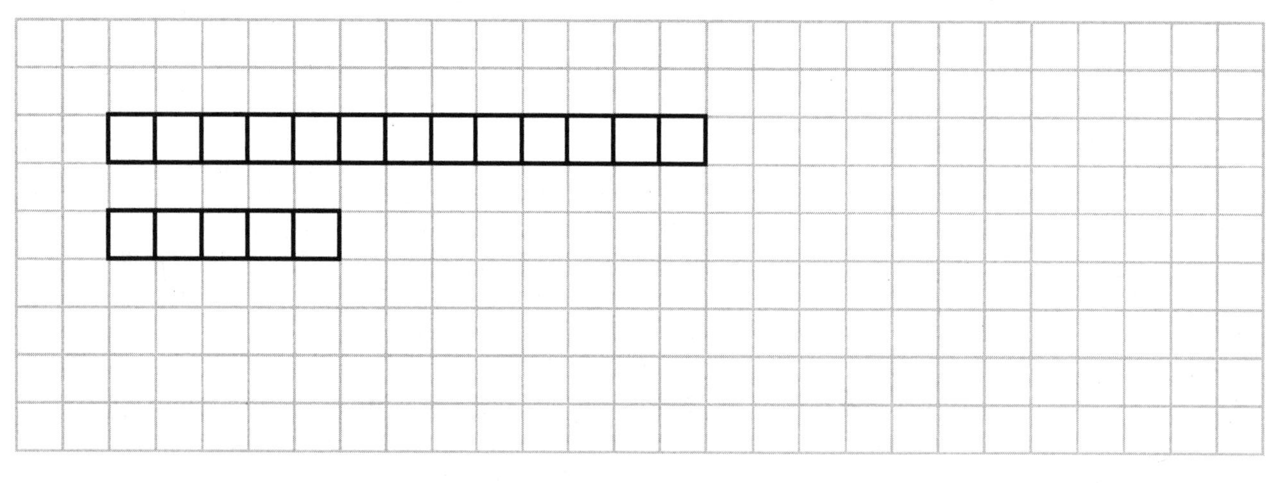

13 + 5 = ______

They made ______ purple pancakes altogether.

Name ______________________________

35. Cat R. Pillar has 10 socks. He found 8 more under the bed. How many socks does Mr. Pillar have now?

$10 + 8 =$ ______

Mr. Pillar has ______ socks now.

36. Moe is 8 years old now. How old will he be in 12 years?

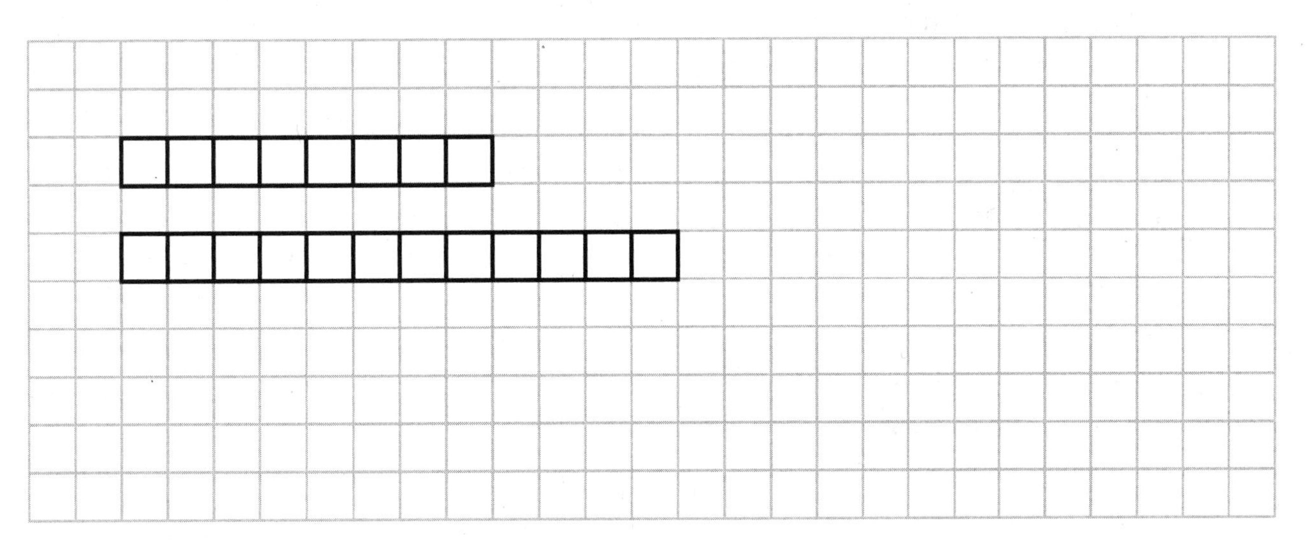

$8 + 12 =$ ______

Moe will be ______ years old.

Name ______________________________

37. Tom and Tim each has 5 bananas. Tip has 6 bananas. How many bananas do the boys have altogether?

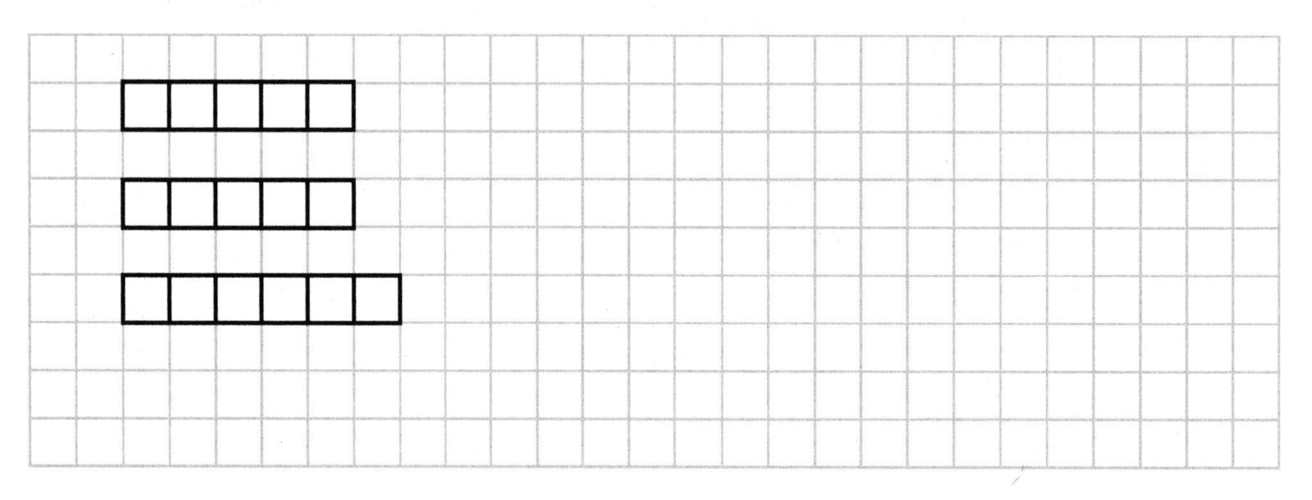

5 + 5 + 6 = ______

The boys have ______ bananas altogether.

38. Dot saw 6 apes. She saw 3 bandicoots. She saw 7 lemurs. How many animals did she see in all?

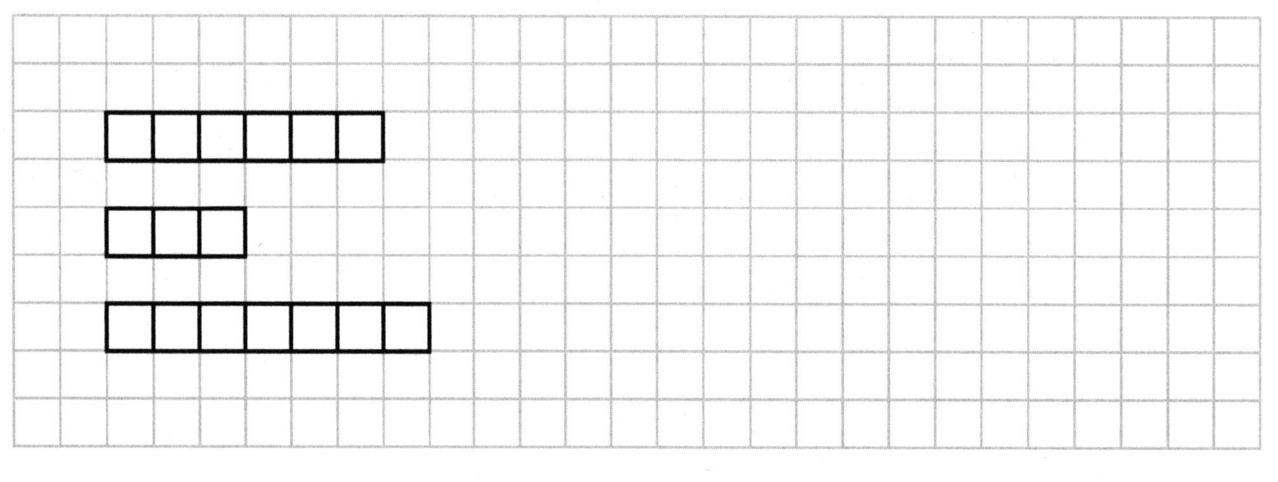

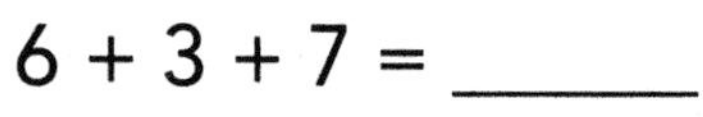

6 + 3 + 7 = ______

Dot saw ______ animals in all.

Name ______________________________

39. Sandy has 7 sand pails. She has 4 sand shovels. She also has 1 sand rake. How many sand tools does Sandy have?

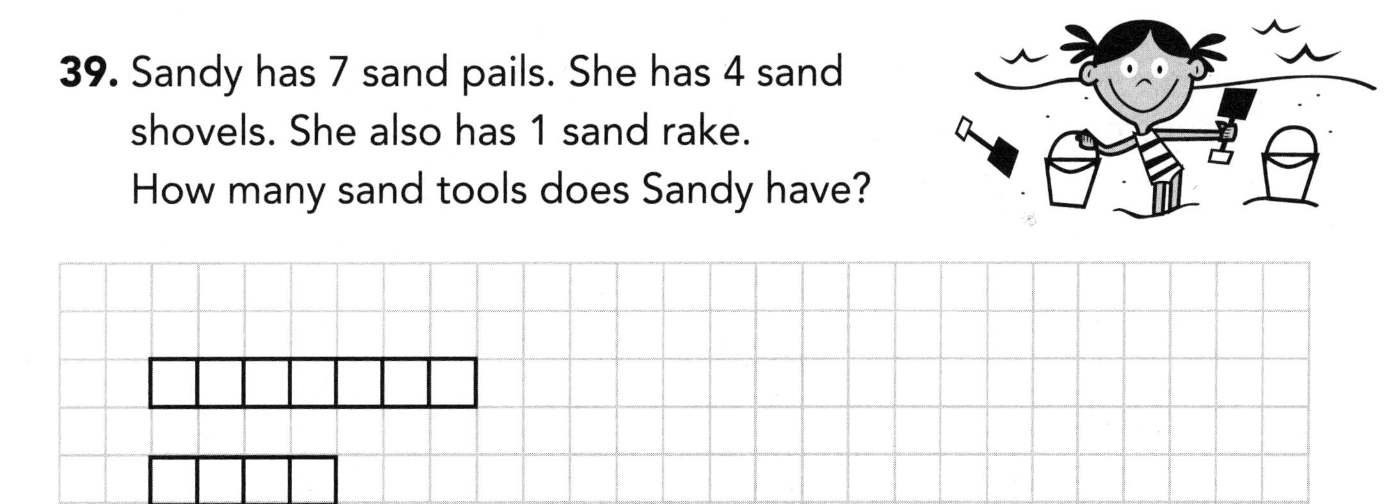

7 + 4 + 1 = ______

Sandy has ______ sand tools.

40. Lucky has 7 black cats and 8 white ones. He has 2 orange ones too. How many cats does Lucky have in all?

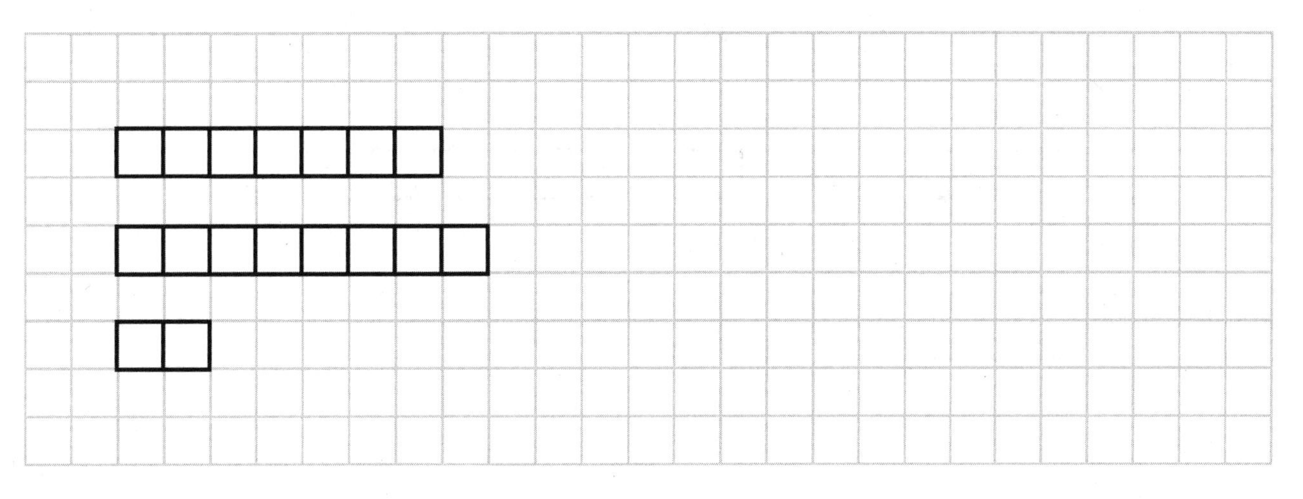

7 + 8 + 2 = ______

Lucky has ______ cats in all.

Name ______________________________

41. There were 15 books on the shelf.
Bingo the bird took 5 of the books.
How many books are left on the shelf?

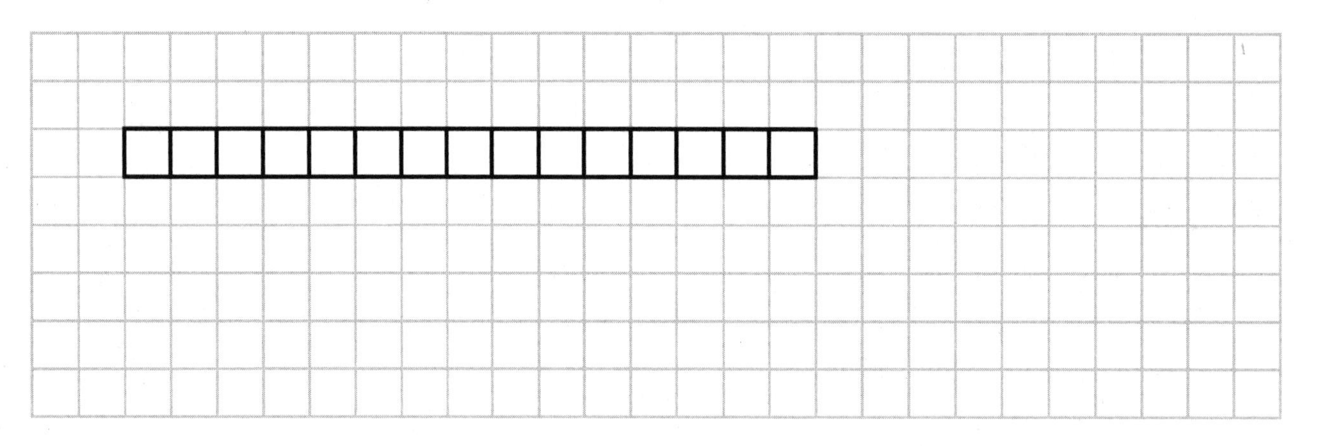

15 – 5 = ______

There are ______ books left on the shelf.

42. Tricky made 18 moose muffins.
Her brother Trucky made 6 moose muffins. How many more moose muffins did Tricky make than Trucky?

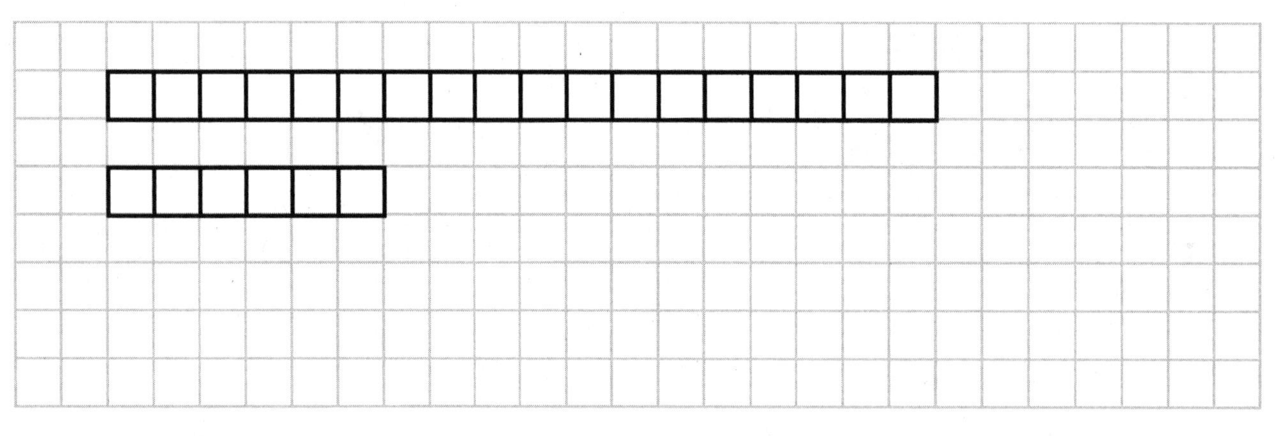

18 – 6 = ______

Tricky made ______ more moose muffins than Trucky.

Name ____

43. Yan has 17 hats.
He put feathers on 10 of them.
How many hats do not have feathers?

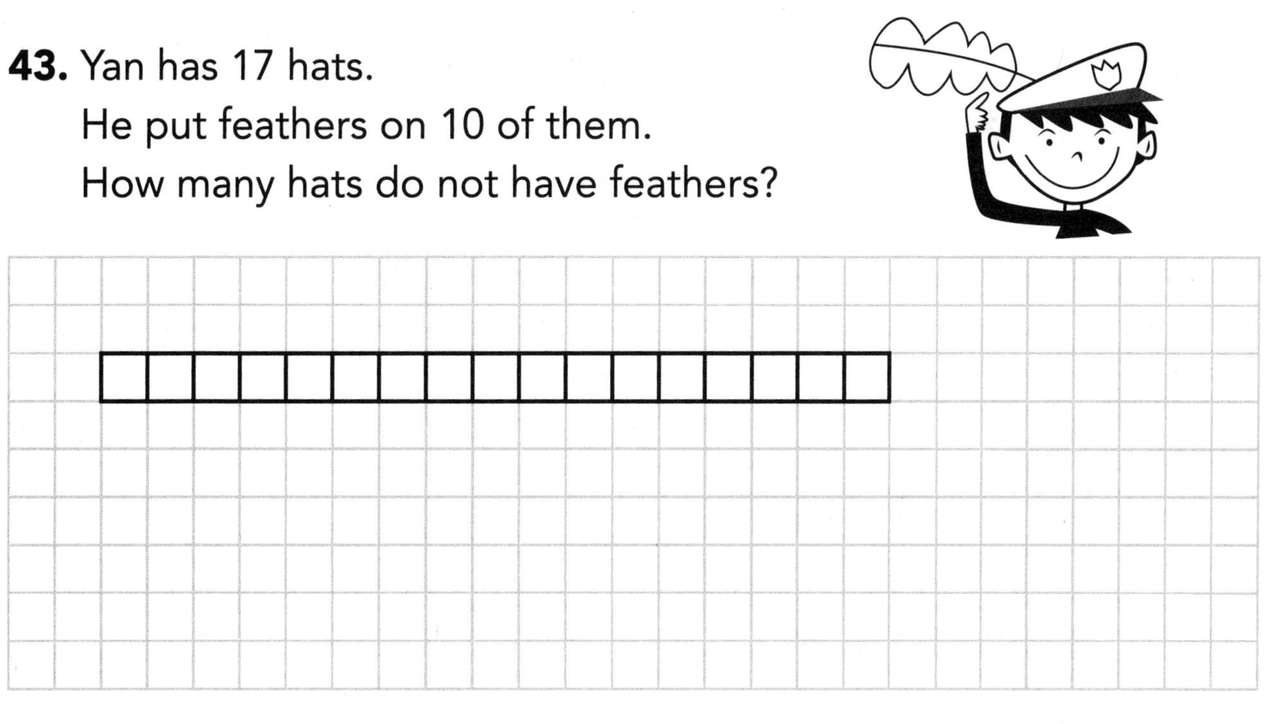

17 – 10 = ______

______ hats do not have feathers.

44. Sara had 16 tons of candy. She gave away 8 tons.
How many tons of candy does she have left?

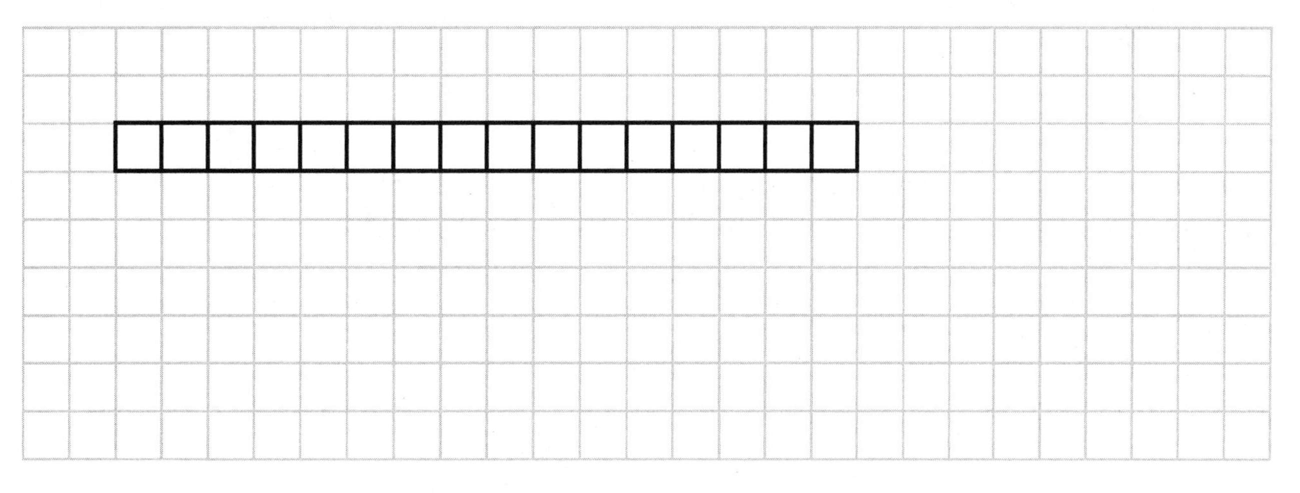

16 – 8 = ______

Sara has ______ tons of candy left.

Name ___________________________________

45. There were 20 boats at the dock.
10 of the boats sailed to Boo Boo Island.
How many boats are still at the dock?

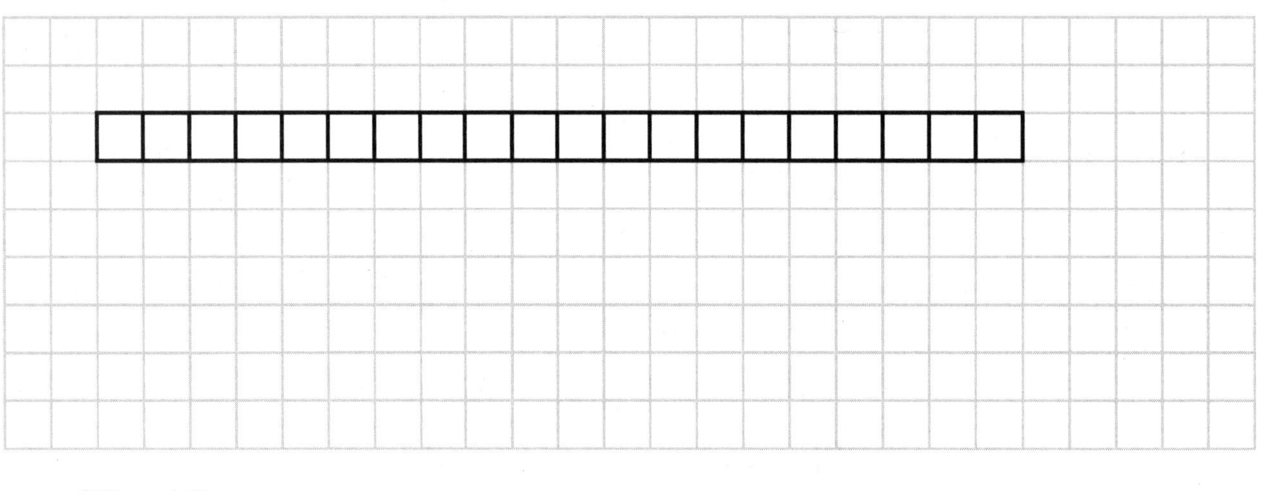

20 – 10 = ______

There are ______ boats still at the dock.

46. There are 18 glasses on the table.
Bossy the cow put milk in 7 of them.
How many glasses are still empty?

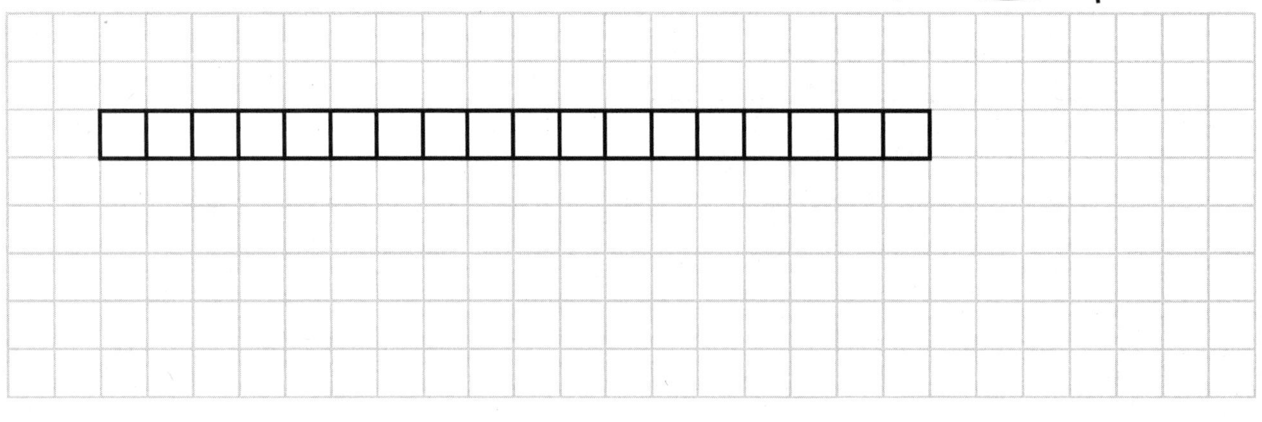

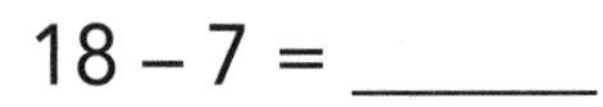
18 – 7 = ______

There are ______ glasses that are still empty.

Name ______________________________

47. There were 14 birds sitting on Stan's head. 9 flew off. How many birds are still on Stan's head?

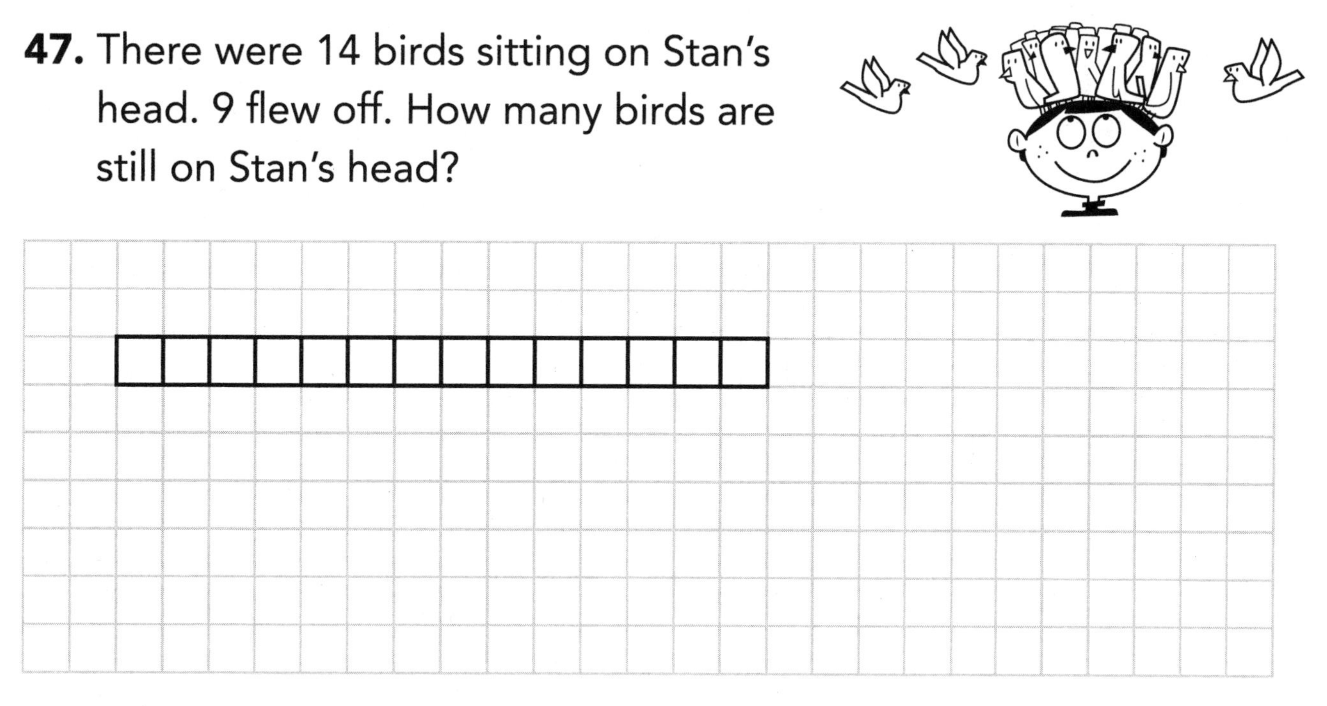

14 – 9 = ______

There are ______ birds still on Stan's head.

48. Zing found 19 diamonds. 6 were blue. The rest were yellow. How many diamonds were yellow?

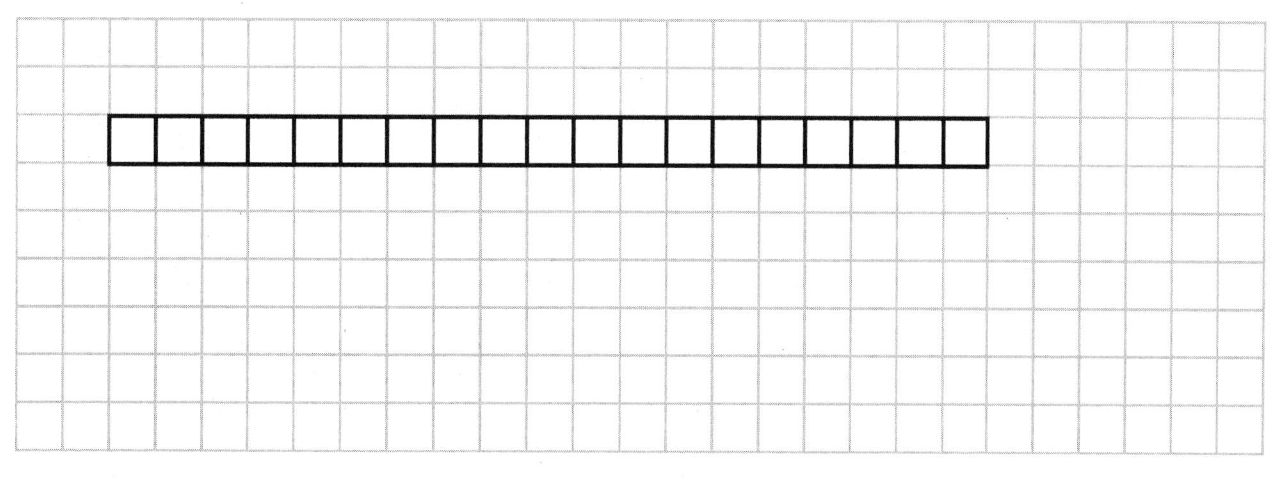

19 – 6 = ______

There were ______ yellow diamonds.

Name ______________________________

49. Slippy Snail crawled 14 inches on Monday.
On Tuesday, he crawled 5 inches.
How many inches did he crawl altogether?

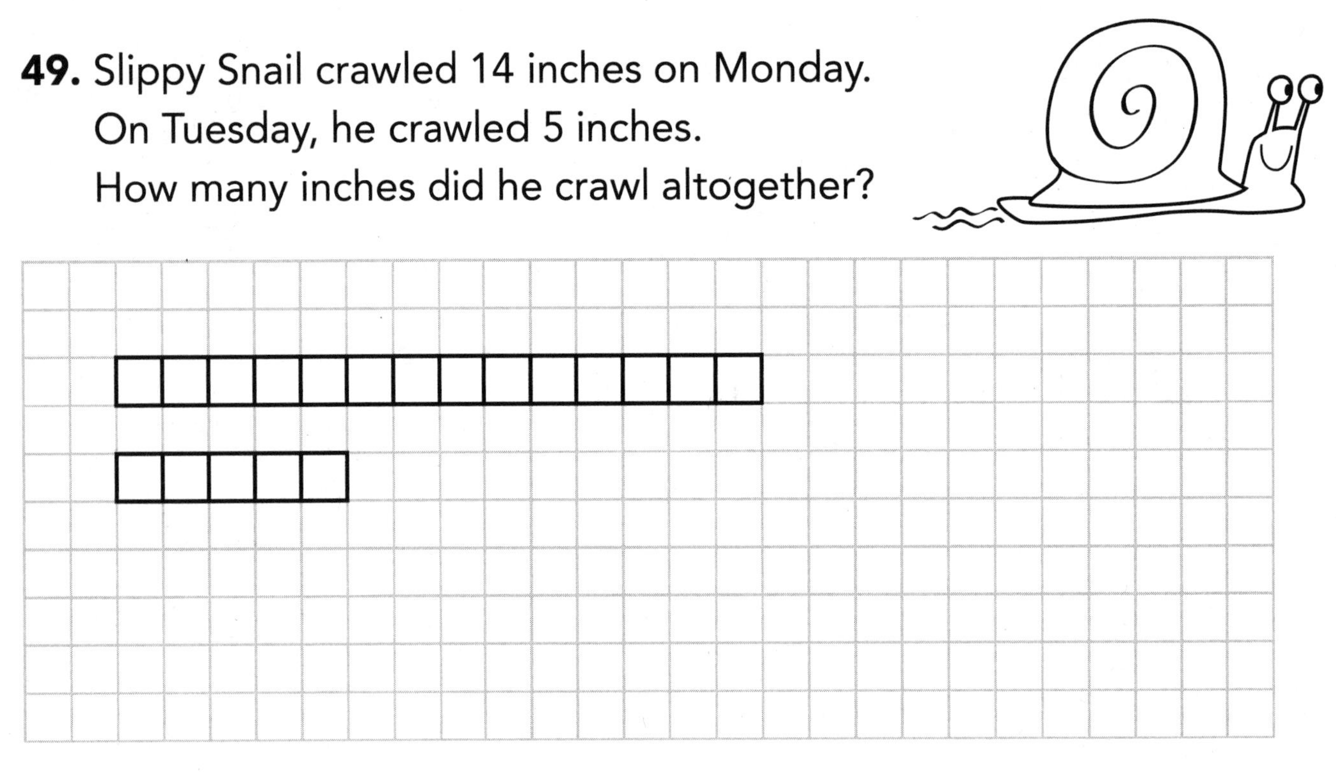

Slippy crawled _____ inches altogether.

50. Oogie wanted to make fruit salad. So he got 6 pounds of grapes, 6 pounds of blueberries, and 6 pounds of kookoo berries.
How much did the fruit salad weigh?

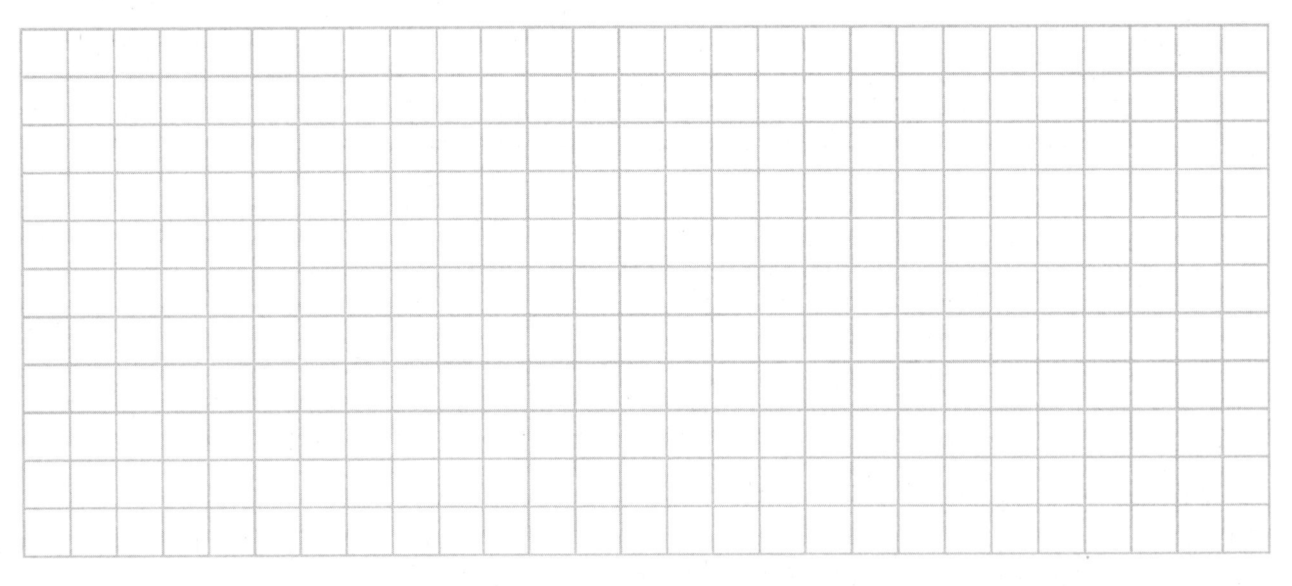

The fruit salad weighed _____ pounds.

Name ____

51. Ray made a fishing pole that was 3 feet long. His friend, George the Giant, made a fishing pole that was 17 feet long. They put their poles together. How long is the new pole?

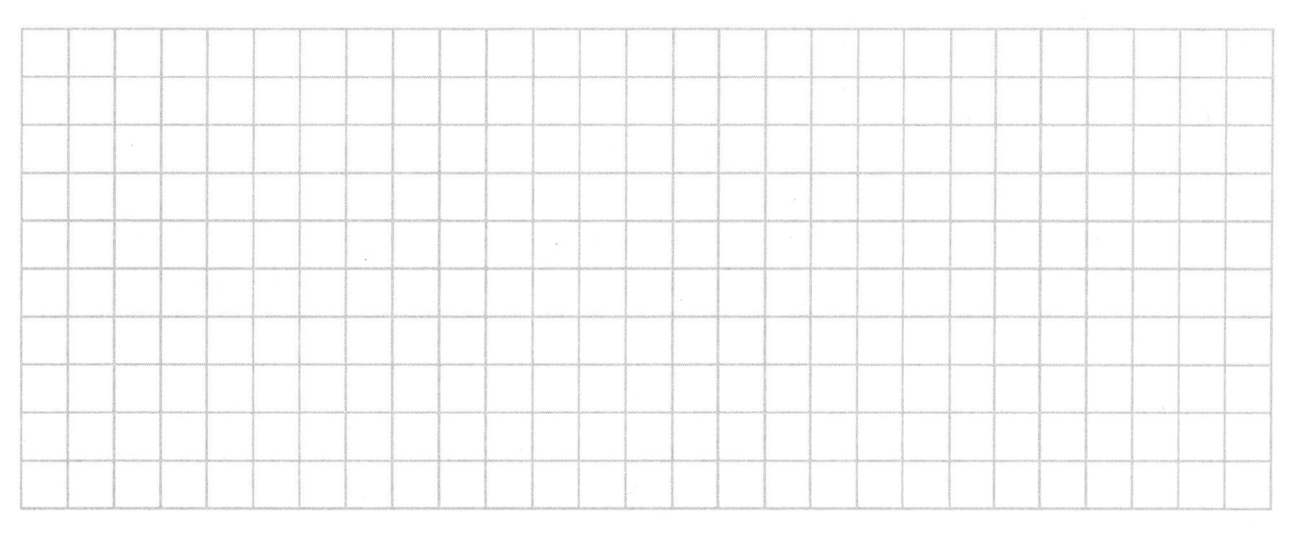

The new pole is ______ feet long.

52. Addie is going to paint her pants. She has 12 ounces of green paint and 7 ounces of orange paint. How many ounces of paint does she have altogether?

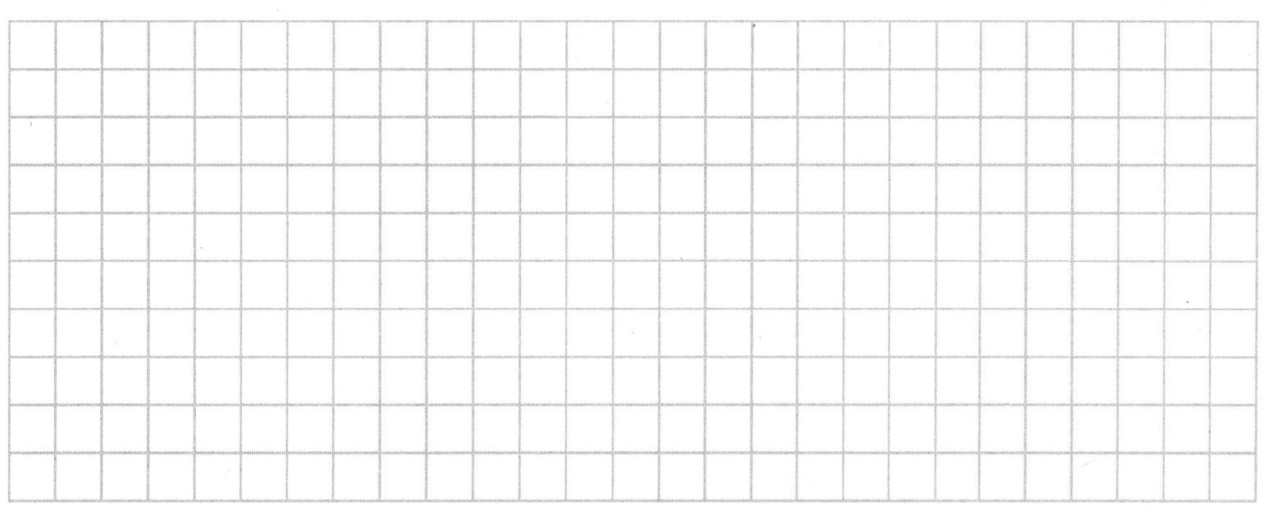

Addie has ______ ounces of paint altogether.

Name ______________________________

53. Kimmy had 9 feet of yarn to make a necklace for an elephant. She cut off 6 feet. How many feet of yarn were left?

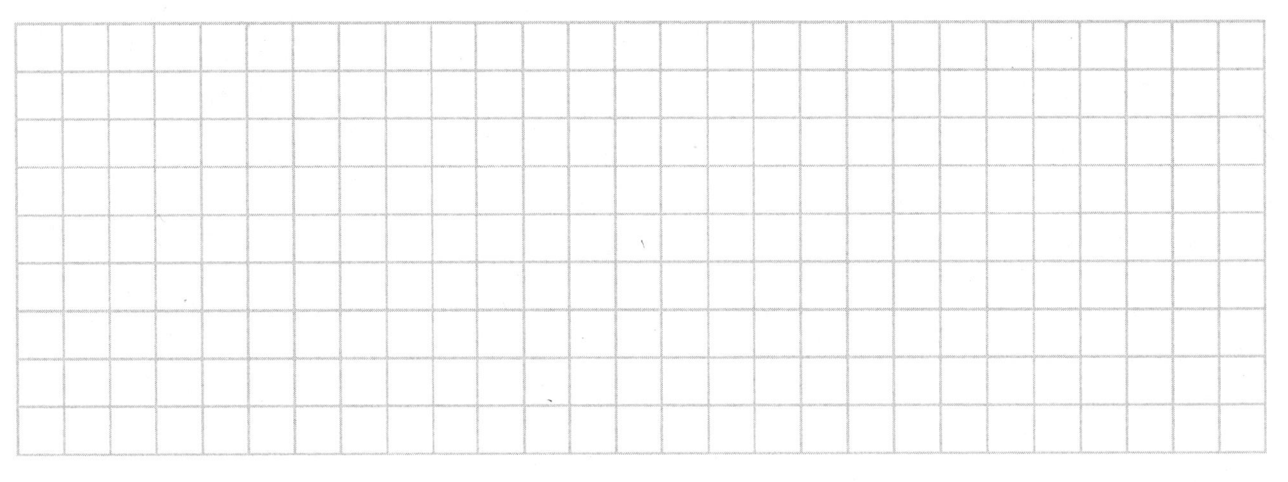

There were ______ feet of yarn left.

54. Jimmy has a rubber-band ball that weighs 19 pounds. Timmy has a rubber-band ball that weighs 6 pounds. What is the difference between the two balls' weights?

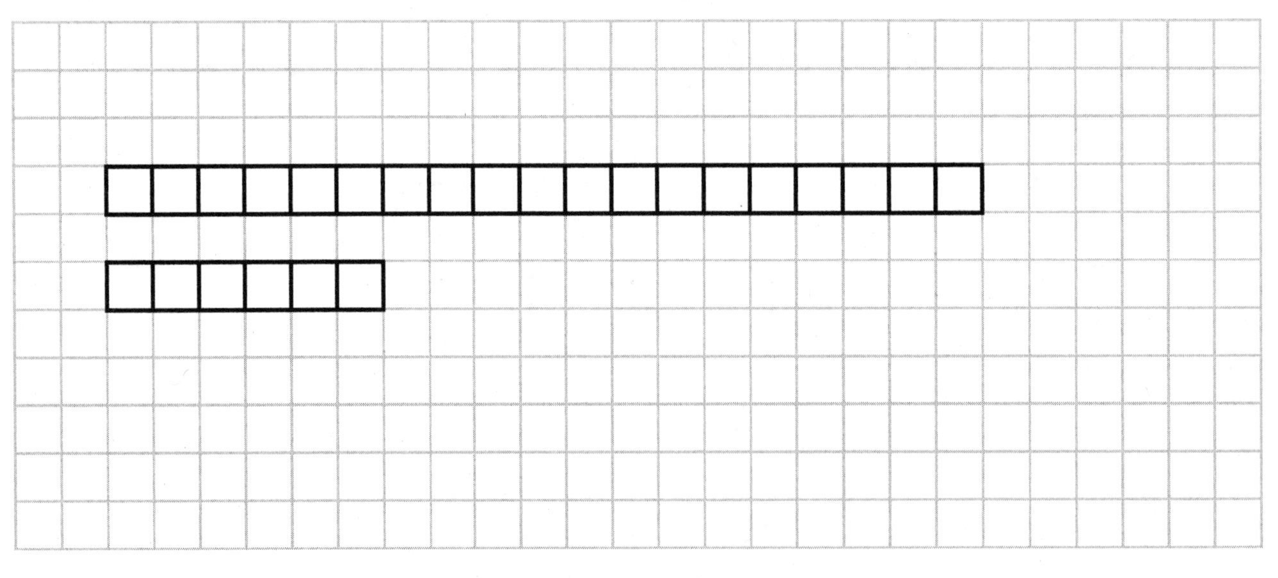

The difference between the two balls is ______ pounds.

Name ____

55. Count Blockula had a black block that was 20 inches high.
He painted 5 inches of it red.
How much of the block was still black?

______ inches of the block were still black.

56. It is 17 feet from Todd's bed to the igloo.
He walked 13 feet. How much farther
does he have to go to get to the igloo?

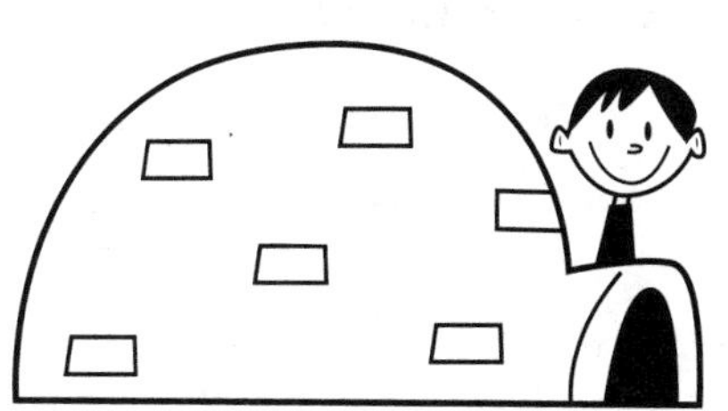

Todd has to walk ______ more feet.

Name ____

57. Gert had 4 dollars. She got 8 dollars for her birthday. How much money does Gert have now?

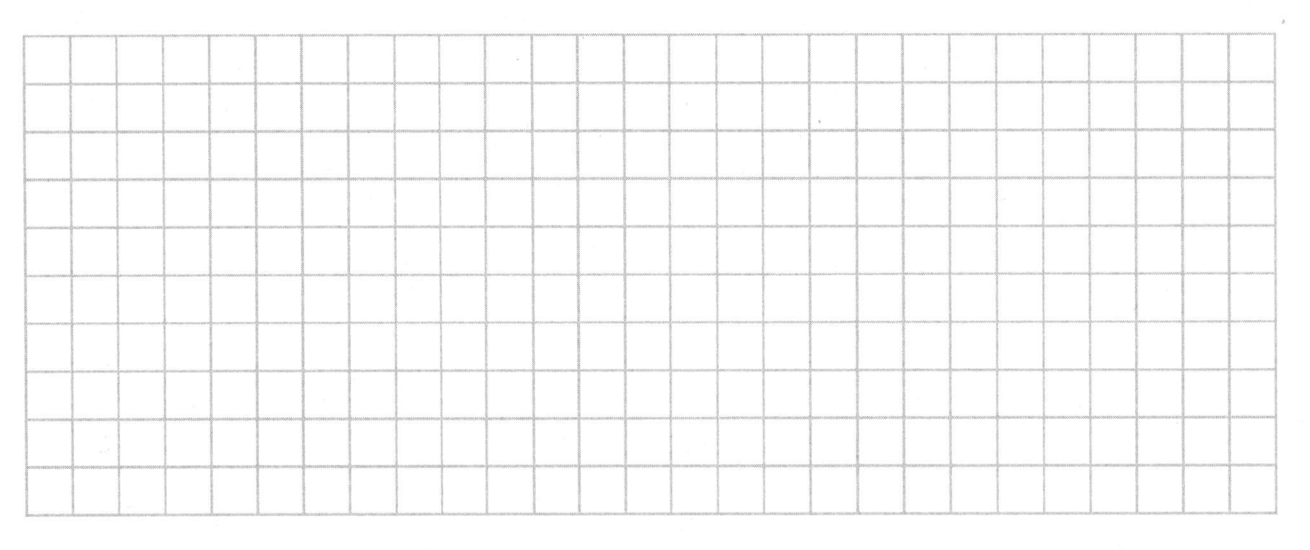

Gert has ______ dollars now.

58. Malik found 4 cents in his sock. He found 4 more cents in his shoe and 5 cents in his ear. How much money did Malik find altogether?

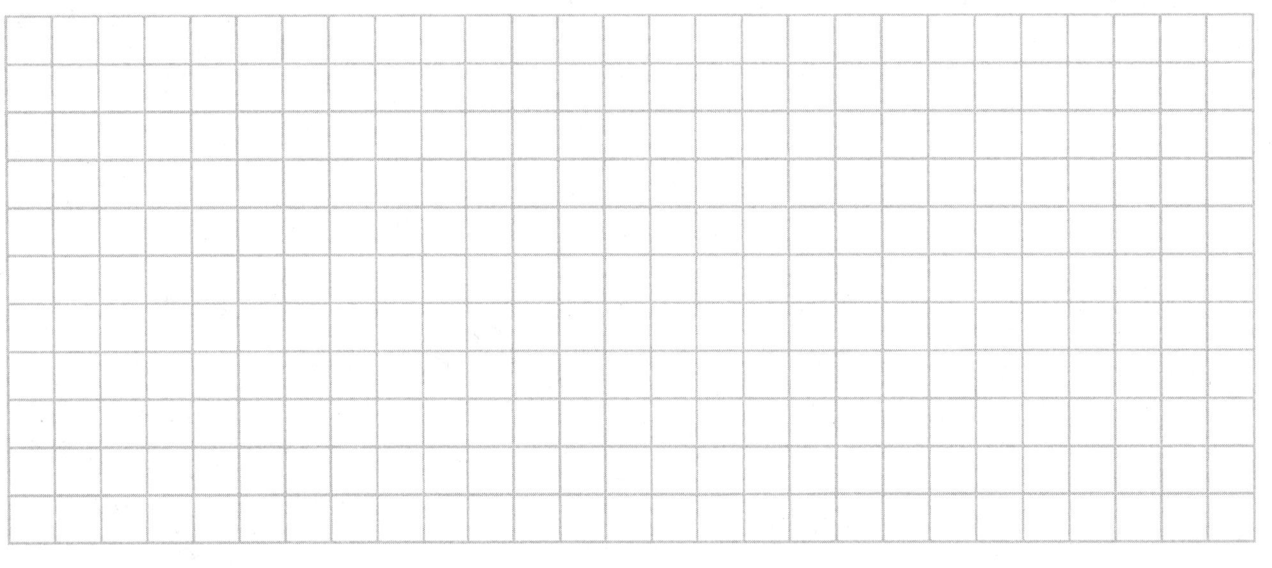

Malik found ______ cents altogether.

Name ______________________________

59. Tamara made $6 walking the cat and $7 walking the rat. How much money did she make in all?

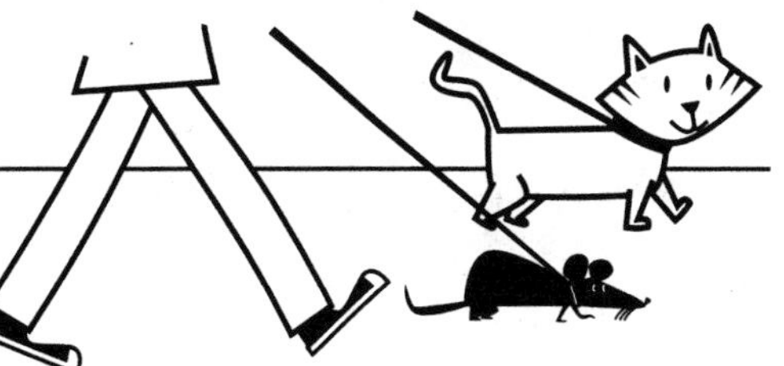

Tamara made $______ in all.

60. Bill had 16 cents. He sold a robot sticker for 4 cents. How much money does Bill have now?

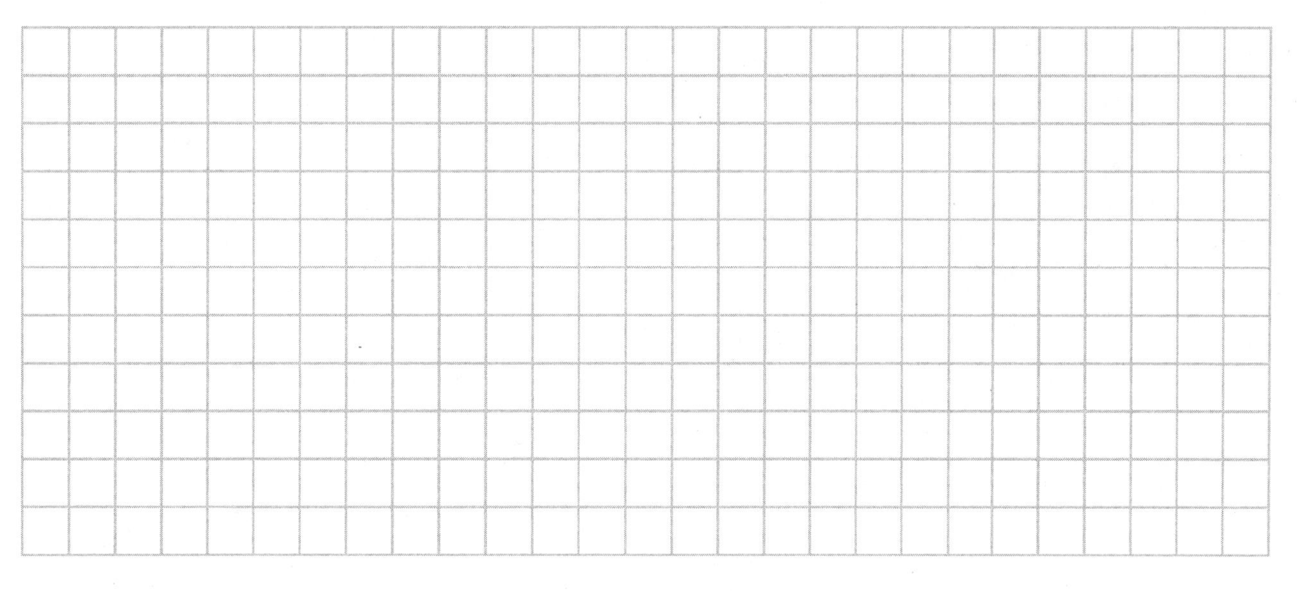

Bill has ______ cents now.

Name ____

61. Uma had 8 dollars. She spent 5 dollars on a movie ticket. How much money does Uma have left?

Uma has ______ dollars left.

62. Hamster has 18 cents. Samster has 6 cents. How much more money does Hamster have than Samster?

Hamster has ______ more cents than Samster.

Name ______________________________

63. Joe got paid 13 dollars. He gave 11 dollars to Jill for an electric apple. How much money does Joe have left?

Joe has ______ dollars left.

64. In the year 2618, a trip to the moon costs $12. A trip back to Earth costs $20. How much more does a trip to Earth cost than a trip to the moon?

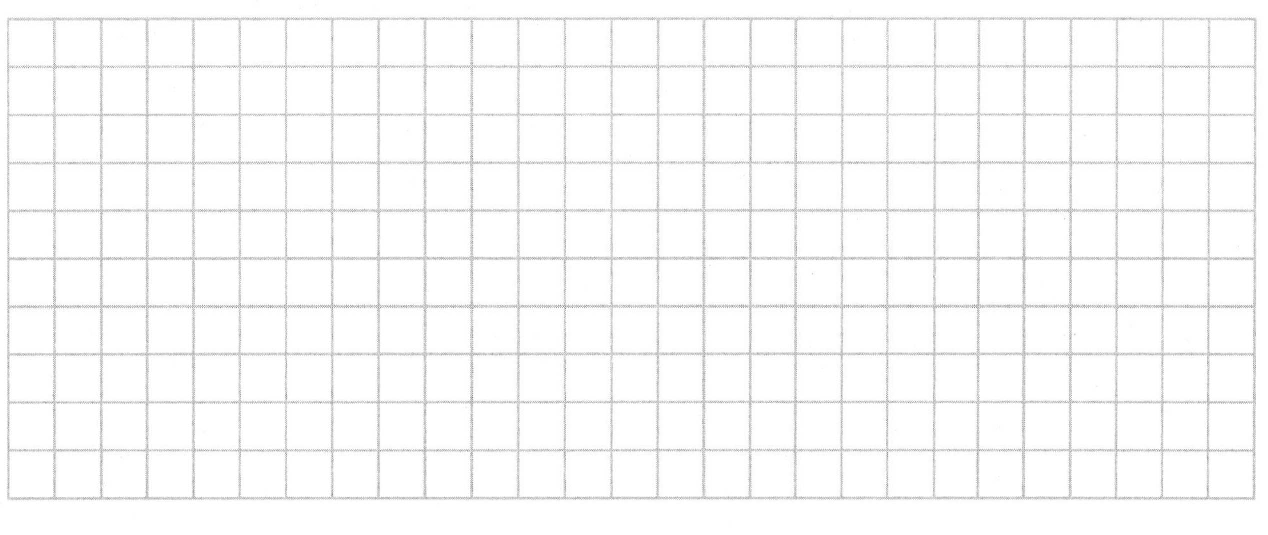

A trip back to Earth costs $______ more than a trip to the moon.

Name ____

65. Nelly Numbers had 4 math books. She bought 3 more.
She gave 5 books to her sister.
How many math books does Nelly have left?

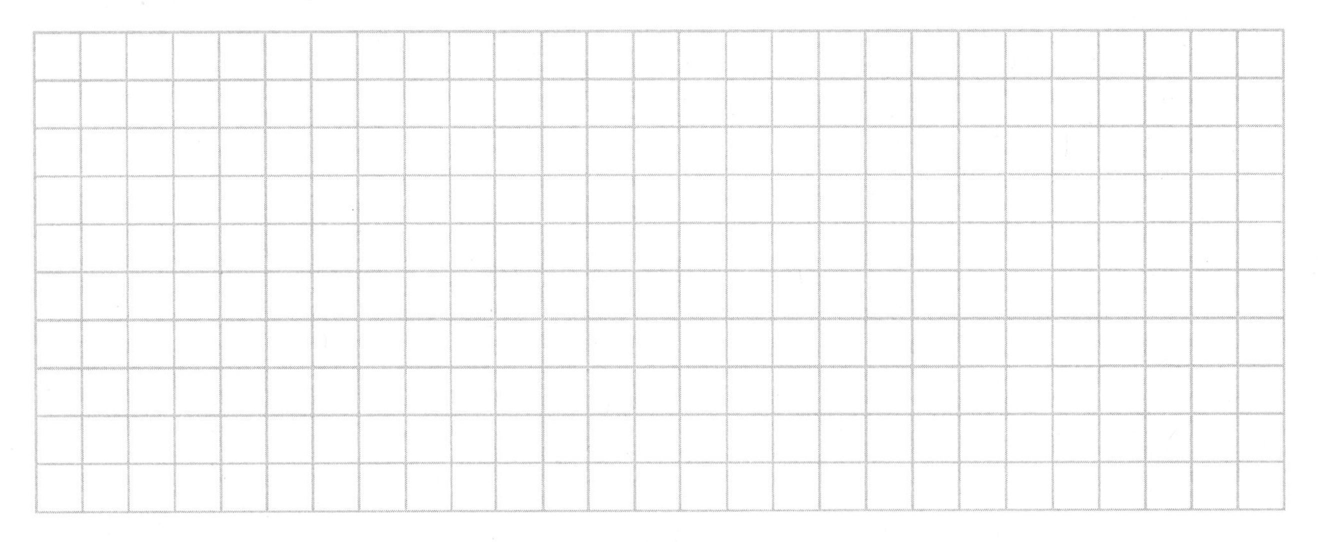

Nelly has ______ books left.

66. King Kong had 8 mini-cupcakes.
He ate 4 of them. Then he bought
5 more later. How many mini-cupcakes
does King Kong have now?

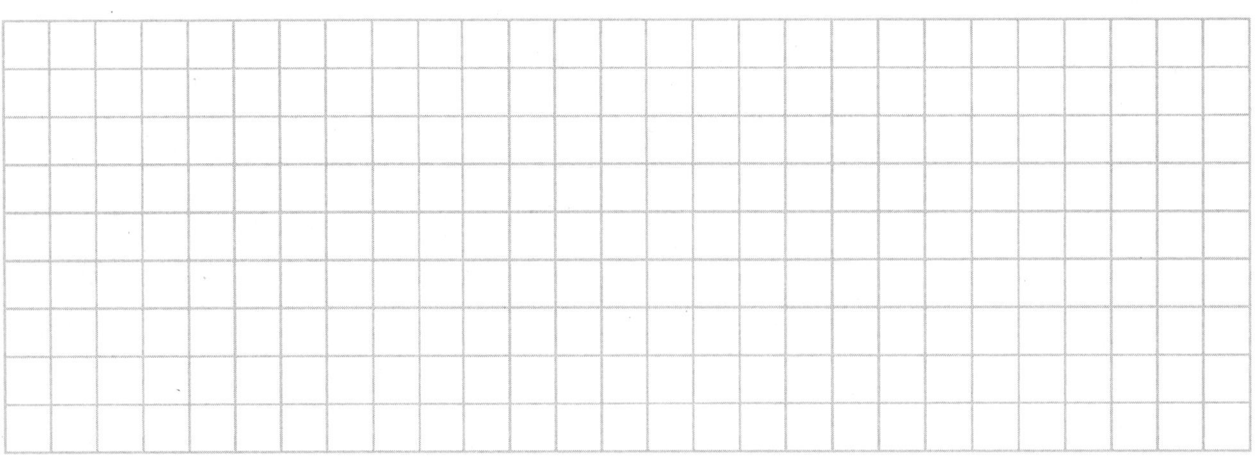

King Kong has ______ mini-cupcakes now.

Name ____

67. There were 12 snails on a rock. 2 more snails joined them. Then 7 snails left. How many snails are still on the rock?

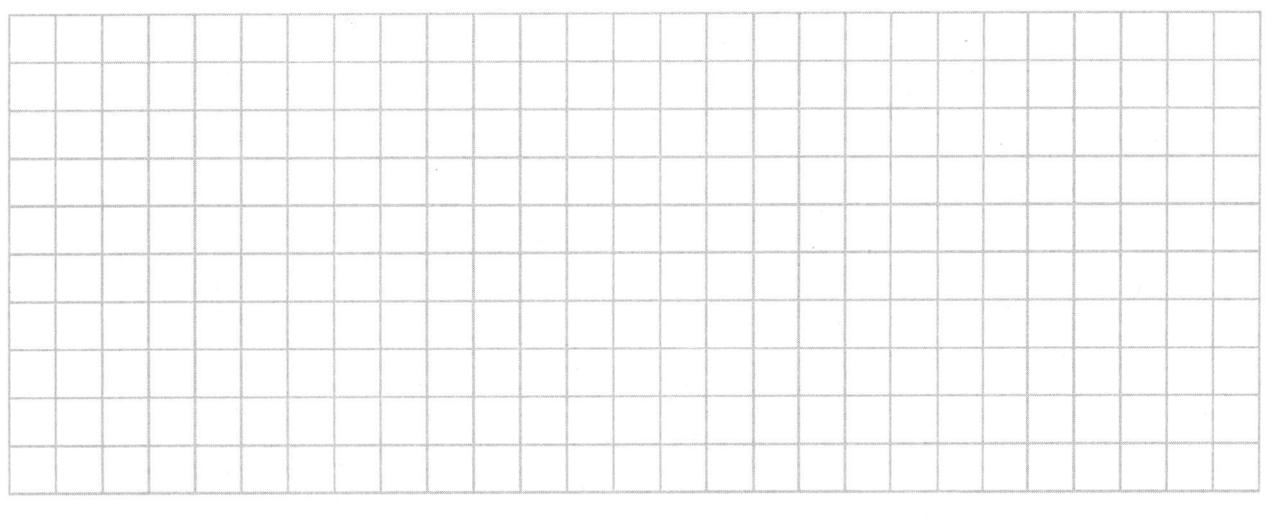

There are ______ snails still on the rock.

68. Pina has 9 bottle caps. Tina has 6 bottle caps. Pina gave Tina 2 bottle caps. Which girl has more bottle caps? How many more?

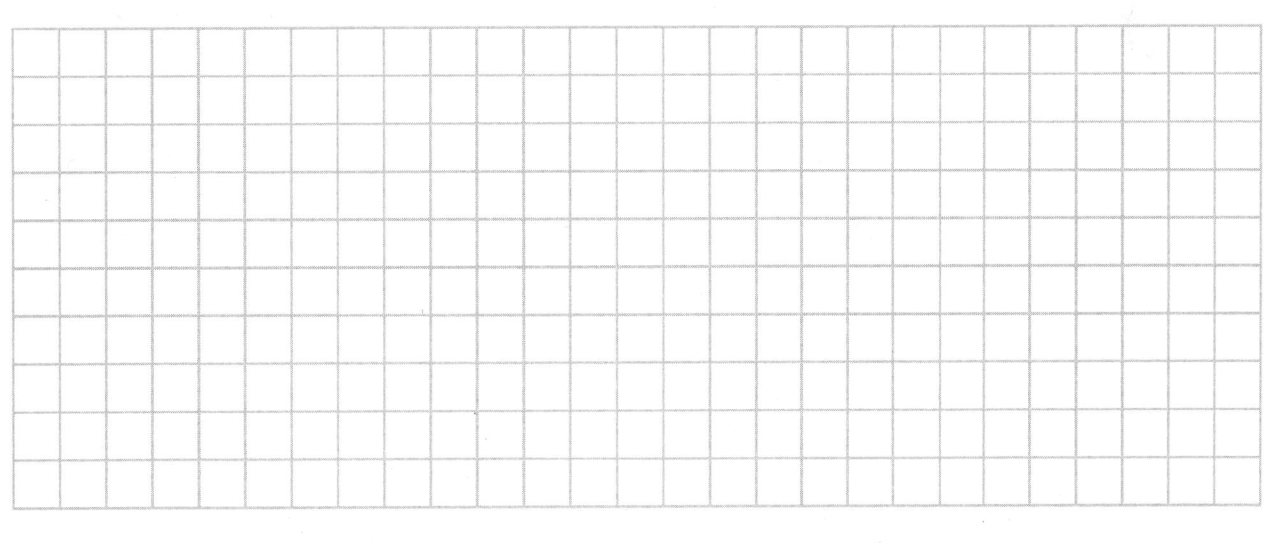

____________ has more bottle caps. She has ______ more.

Name ____

69. Uncle Pete made 7 hot dogs. Uncle Bob made 8 hot dogs. The kids ate 10 of the hot dogs. How many hot dogs were left?

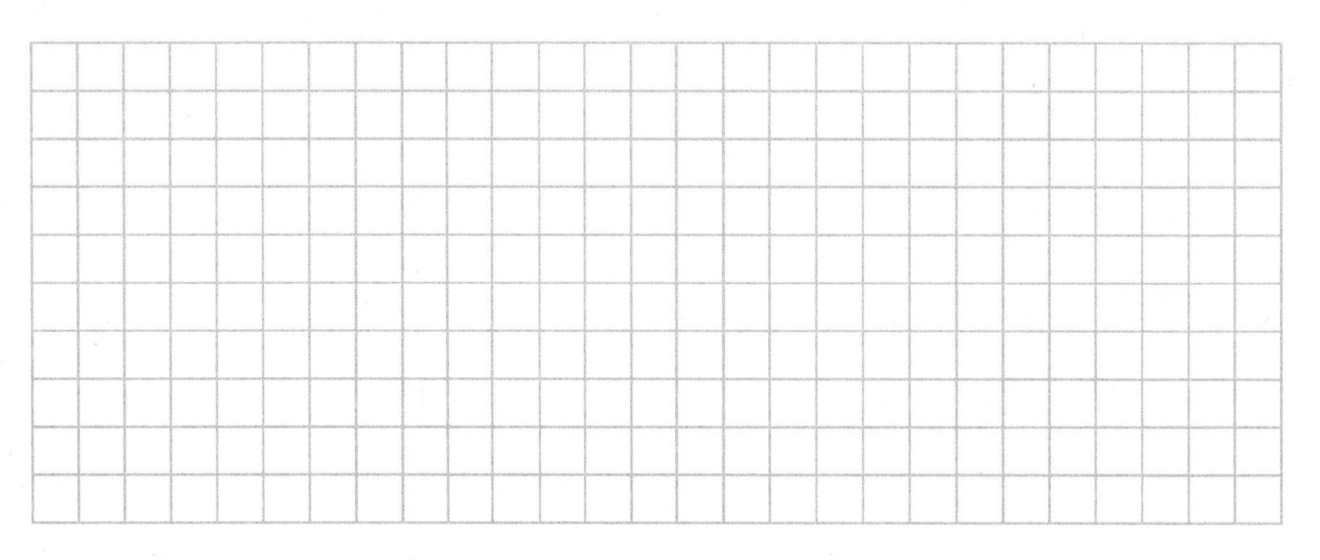

There were ______ hot dogs left.

70. Ringo had 12 rings. He bought 2 more every day for 3 days. How many rings does Ringo have now?

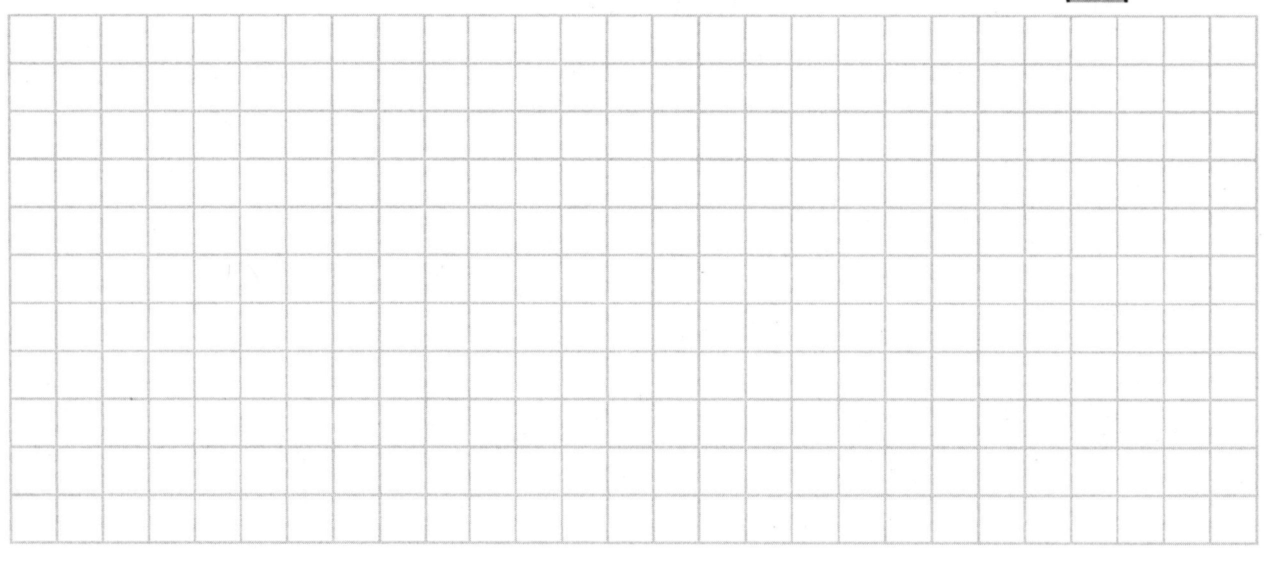

Ringo has ______ rings now.

Name ____

71. Tammy made 17 snowballs. She threw 3 at Pinky. She threw 2 at Dinky. How many snowballs does Tammy have left?

Tammy has ______ snowballs left.

72. Purple the cat had 15 collars. She got 5 more.
Then she gave 8 away.
How many collars does Purple have left?

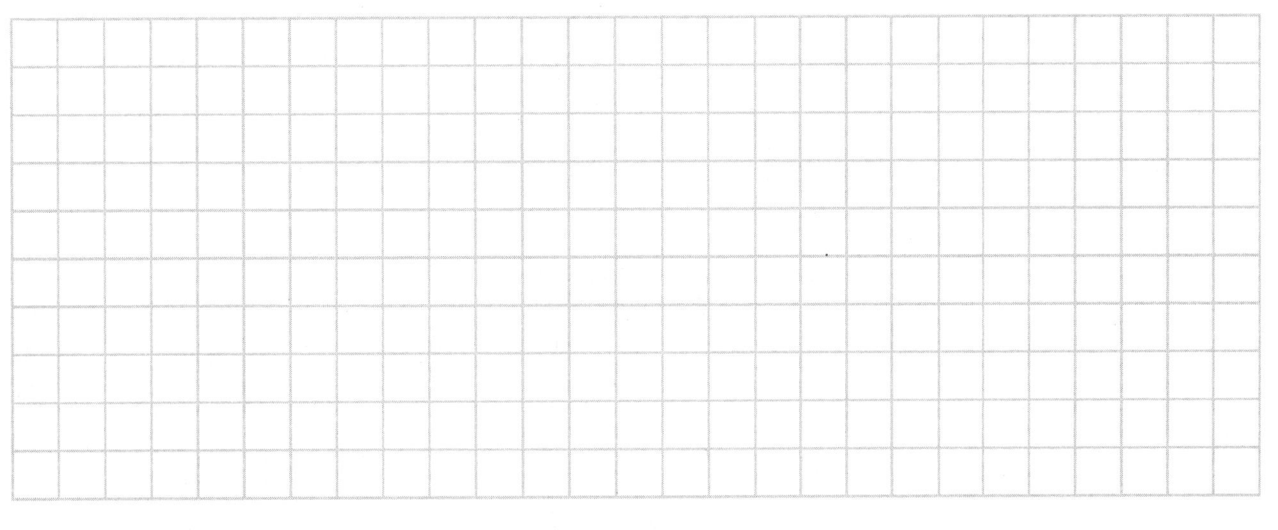

Purple the cat has ______ collars left.

Name ____

73. 25 children came to school wearing hats. 10 hats were yellow. 5 were red. The rest were blue. How many hats were blue?

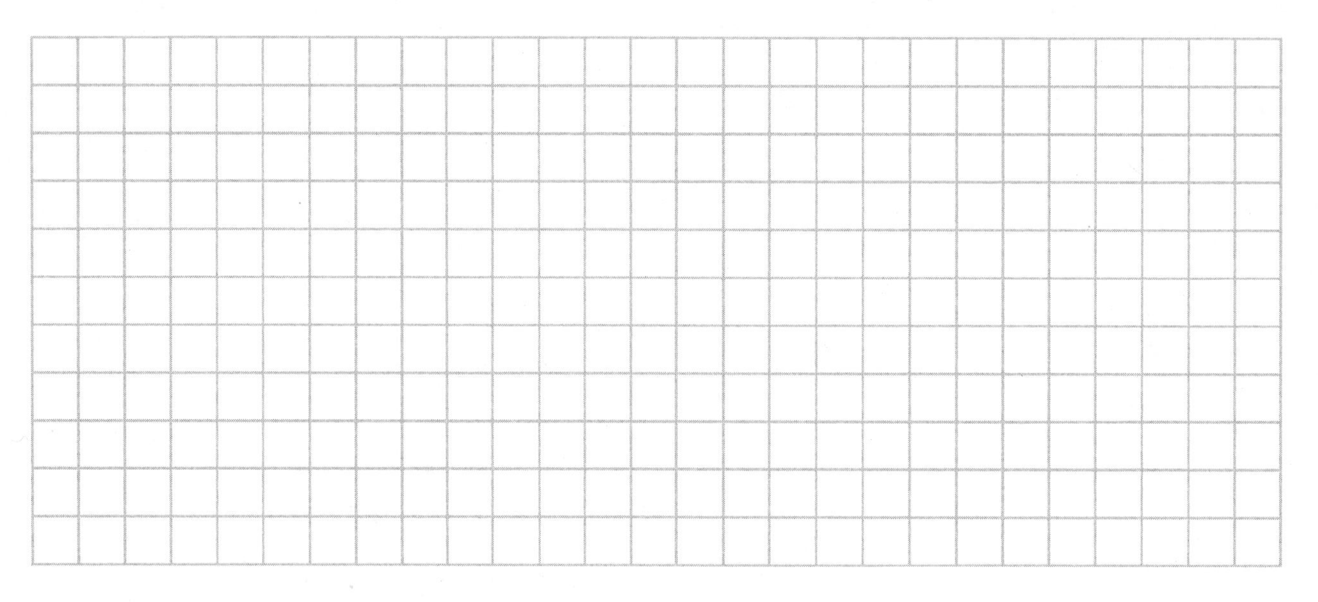

There were ______ blue hats.

74. Doug the Dingo skipped 2 miles every day for 4 days. How many miles did Doug skip altogether?

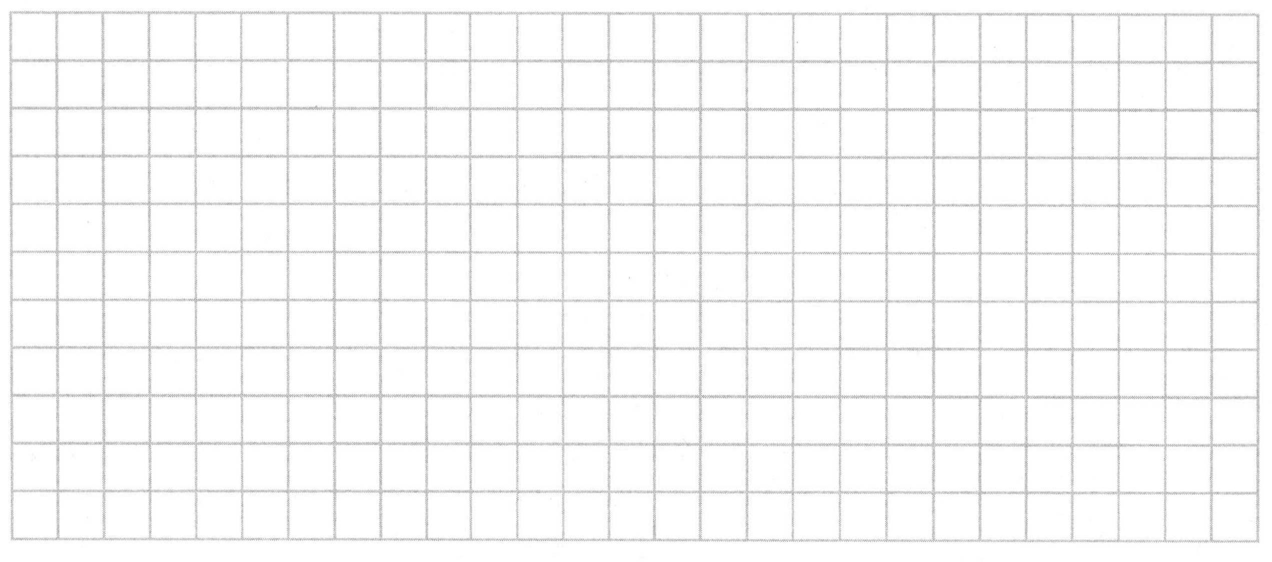

Doug skipped ______ miles altogether.

Name ____

75. Muddy caught 20 fish. Cuddly caught 12 fish on Monday and 6 fish on Tuesday. How many more fish did Muddy catch than Cuddly?

Muddy caught ______ more fish than Cuddly.

76. In the morning, Penrod put 16 napkins on the table for his 10th birthday party. In the afternoon, he put 6 more napkins. 4 guests could not come. So he took 4 napkins away. How many napkins are still on the table?

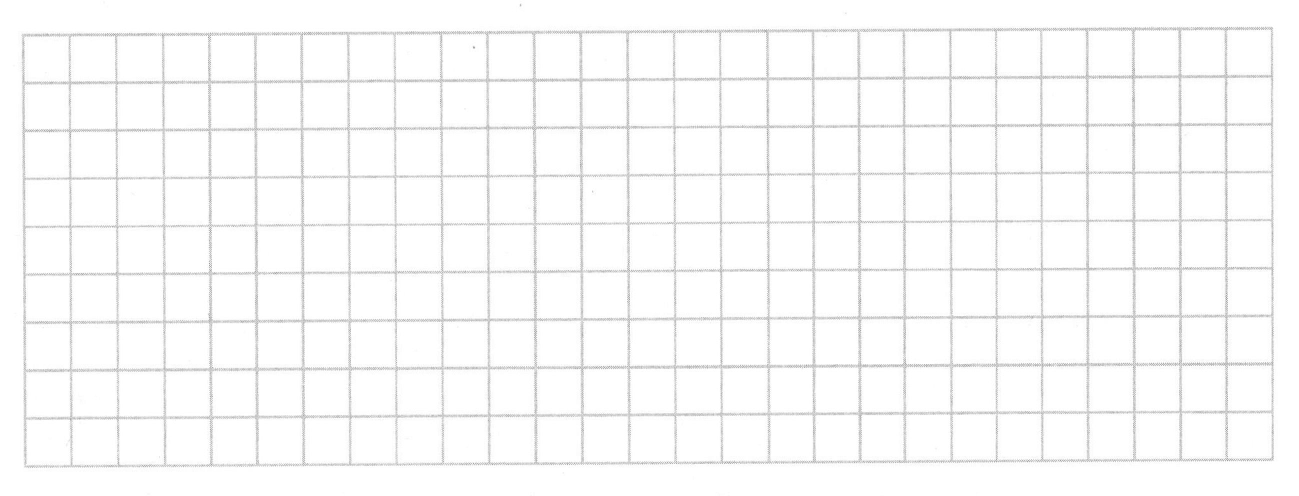

There are ______ napkins still on the table.

Name ________________________________

77. Tangy spent $15 on a new toy mushroom. He only has $19 left. How much money did Tangy start with?

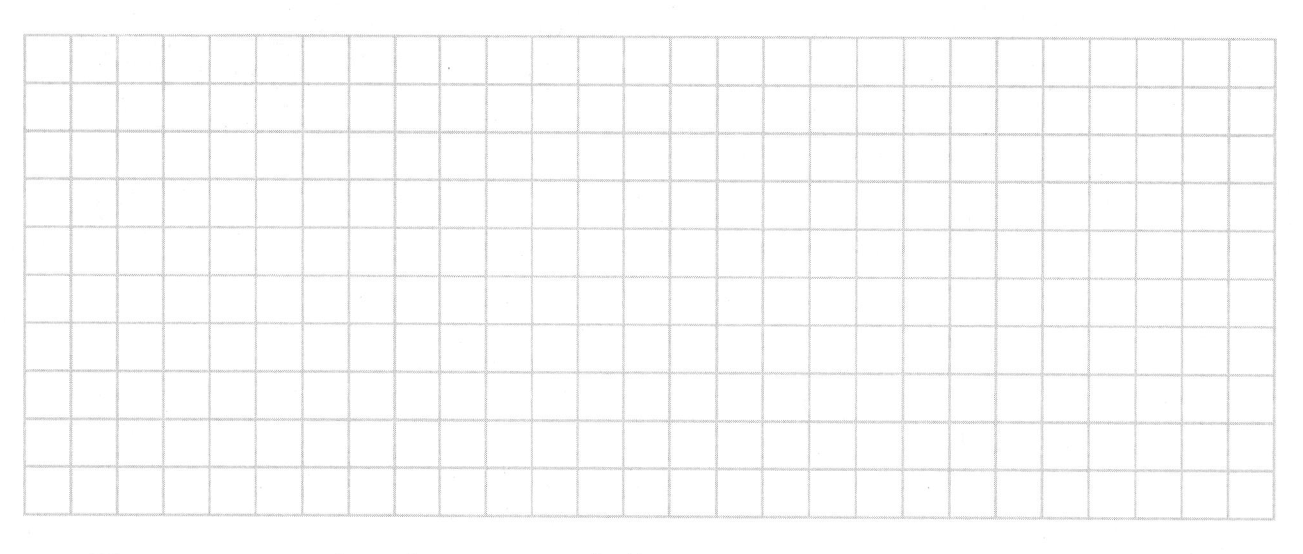

Tangy started with ______ dollars.

78. Lois has three sticks that total 25 feet long. The first stick is 7 feet long. The second stick is 9 feet long. How long is the third stick?

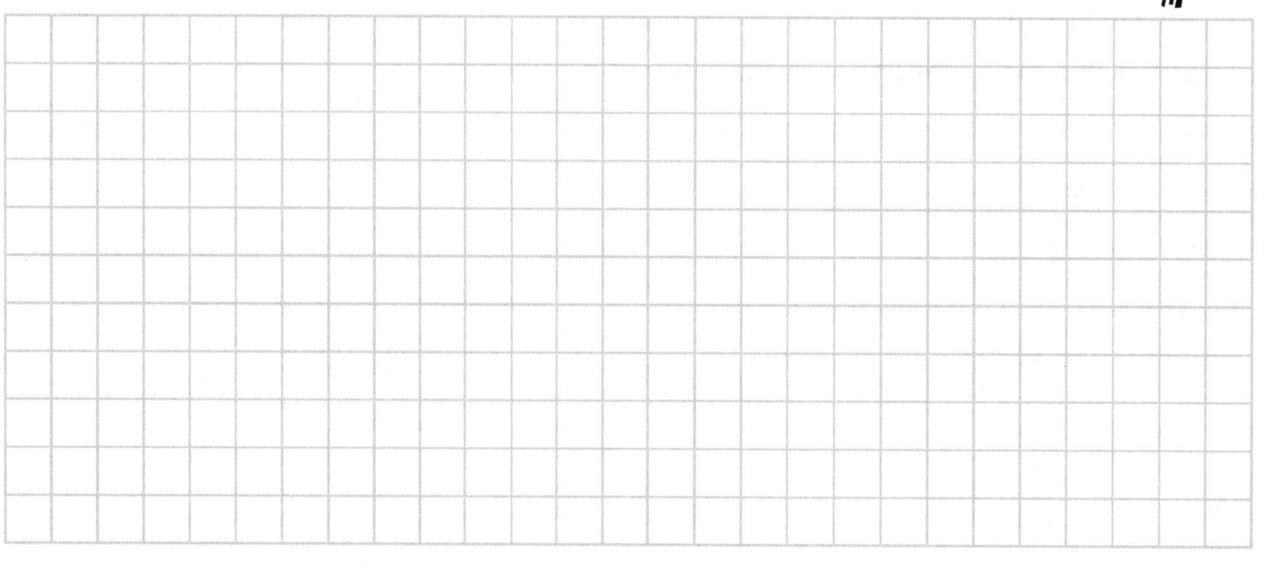

The third stick is ______ feet long.

Name ____

79. Ziggy has 12 black ants and 3 red ants.
Twiggy had 5 black ants and 17 red ants.
Are there more red ants or black ants?
How many more?

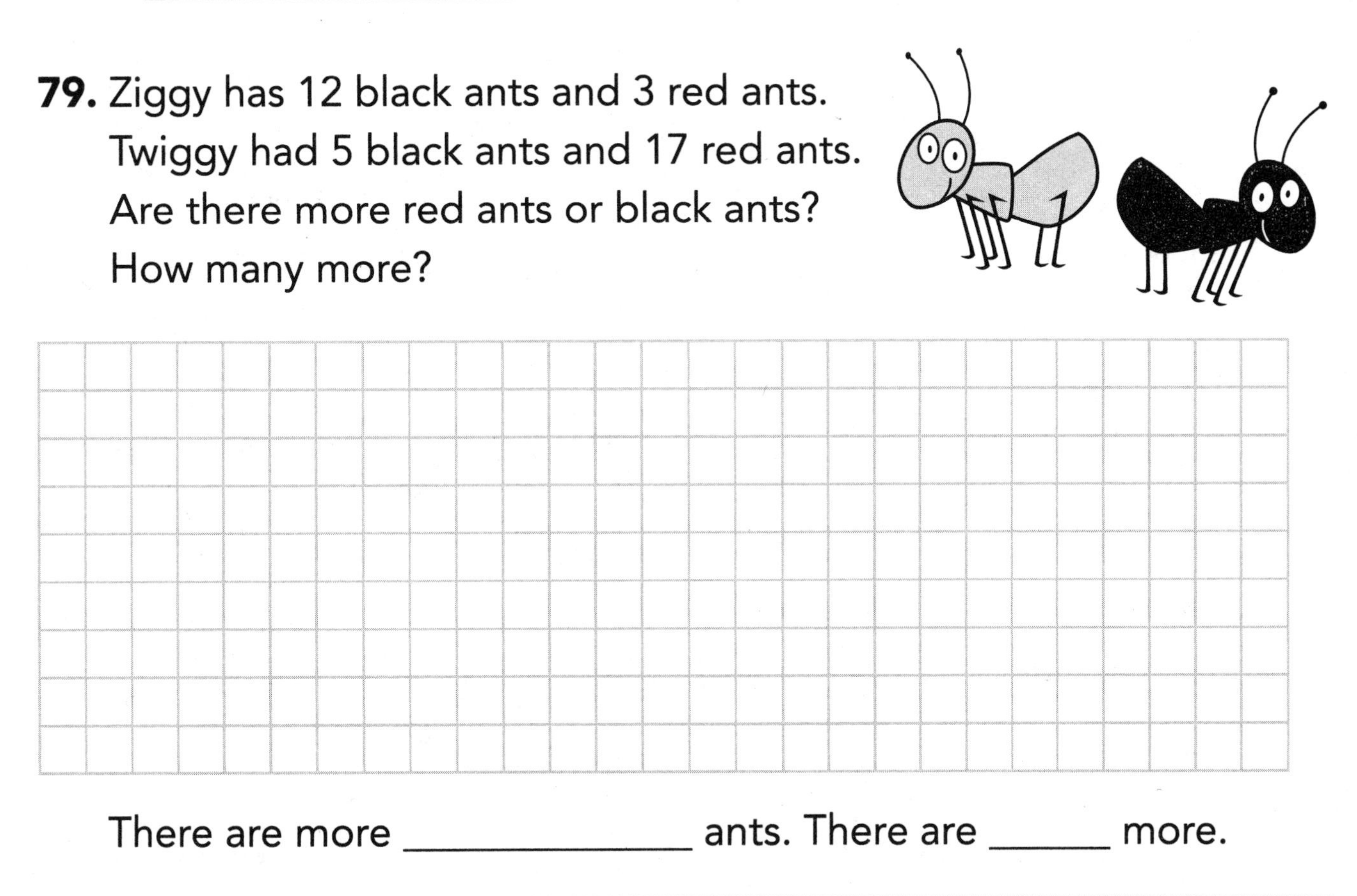

There are more ______________ ants. There are ______ more.

80. There are 80 problems in this book.
Four 1st graders each did 20 problems.
How many problems do they have left to finish the book?

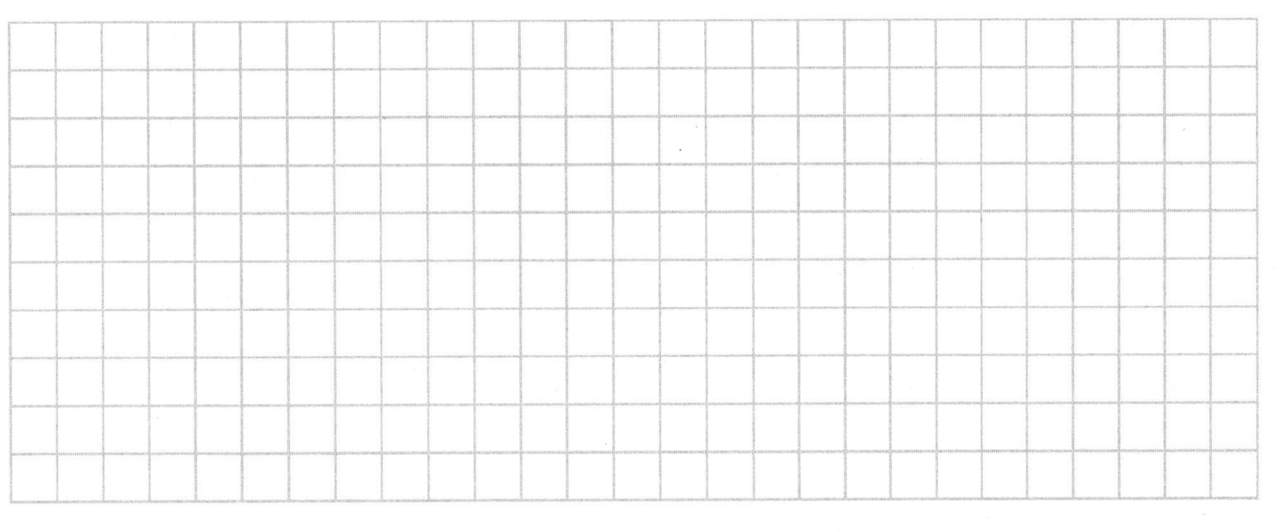

Each 1st grader has ______ problems to do to finish the book.

Answer Key

1. 7 stinkbugs
2. 5 robo-bats
3. 8 bongo berries
4. 4 spaceships
5. 7 super cookies
6. 7 birds
7. 6 coconuts
8. 8 dogs
9. 10 Giganto-Fish
10. 9 birds
11. 10 raccoons
12. 10 hats
13. 9 glow socks
14. 10 dog collars
15. 10 kites
16. 10 bowls
17. 2 moon rocks
18. 5 space rockets
19. 6 noodles
20. 6 buggy bars
21. 2 more dino-hats
22. 6
23. 7 more bonbons
24. 2 more blue pieces
25. 5 doggie cupcakes
26. 8
27. 4 toe rings
28. 7 more blue horns
29. 6 pipsicle sticks
30. 4 more air donuts
31. 5 yellow balls
32. 5 robot bats
33. 13 Zorbo balls
34. 18 purple pancakes
35. 18 socks
36. 20 years old
37. 16 bananas
38. 16 animals
39. 12 sand tools
40. 17 cats
41. 10 books
42. 12 more moose muffins
43. 7 hats
44. 8 tons
45. 10 boats
46. 11 glasses
47. 5 birds
48. 13 yellow diamonds
49. 19 inches
50. 18 pounds
51. 20 feet long
52. 19 ounces
53. 3 feet
54. 13 pounds
55. 15 inches
56. 4 more feet
57. $12
58. 13 cents
59. $13
60. 20 cents
61. $3
62. 12 cents more
63. $2
64. $8 more
65. 2 books
66. 9 mini-cupcakes
67. 7 snails
68. Tina has 1 more bottle cap
69. 5 hot dogs
70. 18 rings
71. 12 snowballs
72. 12 collars
73. 10 blue hats
74. 8 miles
75. 2 more fish
76. 18 napkins
77. $34
78. 9 feet long
79. 3 more red ants
80. 0 problems